Communications in Computer and Information Science 1082

Commenced Publication in 2007
Founding and Former Series Editors:
Phoebe Chen, Alfredo Cuzzocrea, Xiaoyong Du, Orhun Kara, Ting Liu, Krishna M. Sivalingam, Dominik Ślęzak, Takashi Washio, Xiaokang Yang, and Junsong Yuan

More information about this series at http://www.springer.com/series/7899

Deng-Feng Li (Ed.)

Game Theory

Third East Asia International Conference, EAGT 2019
Fuzhou, China, March 7–9, 2019
Revised Selected Papers

Springer

Editor
Deng-Feng Li
School of Economics and Management
Fuzhou University
Fuzhou, Fujian, China

ISSN 1865-0929 ISSN 1865-0937 (electronic)
Communications in Computer and Information Science
ISBN 978-981-15-0656-7 ISBN 978-981-15-0657-4 (eBook)
https://doi.org/10.1007/978-981-15-0657-4

This Springer imprint is published by the registered company Springer Nature Singapore Pte Ltd.
The registered company address is: 152 Beach Road, #21-01/04 Gateway East, Singapore 189721, Singapore

Preface

Recently, non-cooperative and cooperative games, particularly, cooperative games with coalitional structures, fuzzy non-cooperative and cooperative games, dynamic games, evolutionary games, mechanism design, bargaining games, and auctions are attracting significant coverage from the researchers in many subjects or disciplines such as game theory, operations research, mathematics, decision science, management science, economics, experiment economics, system engineering, psychology, and control theory. Also non-cooperative and cooperative games have been successfully applied to various fields such as economics, management, business administration, industrial organization, operations and supply chain management, human resources, energy and resource management, biology, social psychology, and others. In this situation, to strengthen the scientific interaction among the game theory societies in East Asia and to promote academic research, exchange, and collaboration among researchers from East Asia as well as other countries, Fuzhou University of China, East Asia Game Theory Conference Committee, Game Theory Chapter of Operations Research Society of China, Northwestern Polytechnical University of China, Intelligent Decision-Making and Game Chapter of Chinese Society Optimization, Overall Planning and Economical Mathematics hosted the Third International Conference on East Asia Game Theory (EAGT 2019), which was held during March 7–9, 2019, at Fuzhou University, Fujian, China. EAGT is the counterpart of SING (European Meeting on Game Theory) in Asia. EAGT 2019 received 146 abstract submissions and there were more than 180 participants from 13 countries and regions such as Japan, Korea, USA, Russia, Singapore, Italy, the Netherlands, Poland, UK, Canada, India, Hong Kong, and China. Four famous experts and scholars were invited to give keynote talks: Prof. Akihiko Matsui (University of Tokyo, Japan), Prof. Youngsub Chun (Seoul National University, South Korea), Prof. Yukihiko Funaki (Waseda University, Japan), and Prof. Cheng-Zhong Qin (University of California, Santa Barbara, USA).

After EAGT 2019 in Fuzhou, we prepared the proceedings for publication in the Springer series *Communications in Computer and Information Science* (CCIS). Thus, we contacted the experts and scholars who attended EAGT 2019 and invited them to extend their conference papers for consideration in this publication. Finally, we received and accepted nine full papers after two rounds of peer review. The nine papers selected cover topics related to non-cooperative and cooperative games as well as non-cooperative and cooperative games under uncertainty and their applications.

The paper "The Consensus Games for Consensus Economics under the Framework of Blockchain in Fintech," written by Lan Di, Zhe Yang, and George Xianzhi Yuan, introduced a new notion called Consensus Game (CG), motivated by the mechanism design of blockchain economy under the consensus incentives from Bitcoin ecosystems in financial technology (Fintech), the authors established the general existence results for consensus equilibria of consensus games in terms of corresponding interpretation, and based on the viewpoint of Blockchain consensus in Fintech, then applied the concept of hybrid solutions in game theory. As applications, the discussion proposed in this paper illustrates some issues and problems on the stability of mining pool-games for miners by applying consensus games shows that the concept of consensus equilibria could be used as a fundamental tool for the study of consensus economics under the framework of Blockchain economy in Fintech.

The paper "Characterizations of the Position Value for Hypergraph Communication Situations," written by Guang Zhang, Erfang Shan, and Shaojian Qu, studied the position value for arbitrary hypergraph communication situations. The position value was first presented by using the Shapley value of the uniform hyperlink game or the k-augmented uniform hyperlink game, which were obtained from a given hypergraph communication situation. These results generalized the non-axiomatic characterization of the position value from communication situations in Kongo (International Journal of Game Theory, 2010, 39: 669–675) to hypergraph communication situations. Based on the non-axiomatic characterizations, they further provided an axiomatic characterization of the position value for arbitrary hypergraph communication situations by employing component efficiency and a new property, named partial balanced conference contributions. Partial balanced conference contributions were developed from balanced link contributions in Slikker (International Journal of Game Theory, 2005, 33: 505–514).

The paper "A Class of Social-Shapley Values of Cooperative Games with Graph Structure," written by Hui Yang, Hao Sun, and Genjiu Xu, proposed a class of social-Shapley values for cooperative games with graph structure. The social-Shapley value compromised the utilitarianism of the Shapley value and the egalitarianism of the solidarity value in which the sociality was reflected by the solidarity value. Through defining the corresponding properties in graph-restricted games, the authors axiomatically characterized the social-Shapley value when the coefficient is given exogenously. Moreover, they axiomatized the class of all possible social-Shapley values in the graph-restricted cooperative games.

The paper "The Extension of Combinatorial Solutions for Cooperative Games," written by Jiang-Xia Nan, Li-Xiao Wei, and Mao-Jun Zhang, proposed a new model to solve the social selfish coefficient $\alpha \in [0,1]$ for the α-egalitarian Shapley values and a new convex combinations of single-value solutions in terms of a coalition forming weight coefficient $\beta \in [0,1]$. The obtained solution is called the SCE value for cooperative games. The efficiency, linearity, symmetry, and α-dummy player property of the SCE value are proved. By proposing a procedural interpretation, they defined the parameter β as the coalition forming weight (or possibility) coefficient, and found a new way of allocating the grand coalition profit among all players, which coincides with the SCE value. As a result, they verified the validity, applicability, and superiority of the SCE value.

The paper "An Allocation Value of Cooperative Games with Communication Structure and Intuitionistic Fuzzy Coalitions," written by Jie Yang, investigated the cooperative games mainly based on the hypothesis that arbitrary coalitions can be formed and the fuzzy coalitions are Aubin's form. This paper defined a cooperative game with a communication structure and intuitionistic fuzzy coalitions, in which the partners had some degrees of hesitation and different risk preferences when they took part in limited coalitions. There are lower and upper participation degrees of players in coalitions by introducing confidence levels to intuitionistic fuzzy coalitions. Then a formula of an average tree solution (shortly called AT solution) for this cooperative game was proposed based on the defined preference weighted form by taking into account the players' risk preferences, and the existence of the AT solution was proven according to axioms system.

The paper "Shapley Value Method for Benefit Distribution of Technology Innovation in Construction Industry with Intuitionistic Fuzzy Coalitions," written by Ting Han, established a fair and an efficient mechanism to benefit distribution for technology innovation in the construction industry. Firstly, the forming mechanism and value creation mechanism were analyzed. Then, the benefit distribution under the condition that members have a certain degree of participation and a certain degree of non-participation in coalitions was discussed. The essence is to solve cooperative games with intuitionistic fuzzy coalitions, which are called intuitionistic fuzzy cooperative games for short. The Shapley values for intuitionistic fuzzy cooperative games were proposed by using the intuitionistic fuzzy set theory, Choquet integrals, and continuous ordered weighted average operators. It is proved that the defined Shapley value satisfies three axioms, which are desired for any cooperative games.

The paper "The Existence of the pesu-Ky Fan' Points and the Applications in Multi-objective Games," written by Xiaoling Qiu, proposed pesu-Ky Fan' points for vector Ky Fan inequalities and proved the existence results under some relaxed assumptions by virtue of the KKMF principle and Fan-Browder fixed point theorem. Mild continuity, named pseudo-continuity and introduced for the existence results, is weaker than semi-continuity and generalized the present results in the literature. The author defined pesu-weakly Pareto-Nash equillibria for multi-objective games and obtained some existence theorems.

The paper "XGBoost-driven Harsanyi Transformation and Its Application in Incomplete Information Internet Loan Credit Game," written by Yi-Cheng Gong, Yan-Na Zhang, and Li Yu, introduced the statistical learning method known as extreme Gradient Boosting (XGBoost), and hereby proposed an XGBoost-driven Harsanyi transformation, where the XGBoost was used to predict a new player's type distribution indirectly. To test the effect of the XGBoost-driven Harsanyi transformation, an incomplete information Internet loan credit game (3ILCG) was modeled and analyzed. When the loan interest rate was $r = 0.2$, the empirical analysis could be executed on 24,000 training data and 6,000 testing data, then the experiment showed that the accuracy (A) and harmonic mean (F1) of the enterprise loan decision based on p_{xgb} on 6,000 testing data were 0.900833 and 0.945864, respectively. The test experiment demonstrated that the XGBoost-driven Harsanyi transformation can help the lending platform to make loan decisions scientifically in practice and improve the practice value of game theory.

The paper "The Method for Solving Bi-matrix Games with Intuitionistic Fuzzy Set Payoffs," written by Jiang-Xia Nan, Li Zhang, and Deng-Feng Li, developed a bilinear programming method for solving bi-matrix games in which the payoffs are expressed with intuitionistic fuzzy sets (IFSs), which are called IFS bi-matrix games for short. In this method, using the equivalent relation between IFSs and interval-valued fuzzy sets (IVFSs) and the operations of IVFSs, they proposed a new order relation of IFSs through introducing a ranking function, which is proven to be a total order relation.

Hereby they introduced the concept of solutions for IFS bi-matrix games and established parametric bi-matrix games. It was proven that any IFS bi-matrix game has at least one satisfying Nash equilibrium solution, which is equivalent to the Nash equilibrium solution of corresponding parametric bi-matrix game. The latter can be obtained through solving the auxiliary parametric bilinear programming model.

I would like to thank the hard work of the Academic Program Committee and the Organizing Committee of EAGT 2019, as well as all contributors and reviewers, who truly understand the meaning of cooperative games. In particular, I very much appreciate Prof. Dr. Yukihiko Funaki (Waseda University, Japan), Prof. Dr. Xiao-Guang Yang (Academy of Mathematics and Systems Science, Chinese Academy of Sciences, China), and Prof. Dr. Genjiu Xu (Northwestern Polytechnical University, China). At the same time, I would like to thank my PhD students, Ms. Bin-Qian Jiang, Prof. Xia-Jiang Nan, Ms. Xiao-Li Du, and Ms. Ping-Ping Lin for their all efforts, inputs, and excellent work in EAGT 2019 and for editing the publication.

Deng-Feng Li

Organization

Hosts

Fuzhou University, China
East Asia Game Theory Conference Committee
Game Theory Chapter of Operations Research Society of China
Northwestern Polytechnical University, China
Intelligent Decision-Making and Game Chapter of Chinese Society Optimization, Overall Planning and Economical Mathematics, China

Organizing Committee

Chair

Deng-Feng Li	School of Economics and Management, Fuzhou University, China

Organizing Committee Members

Zhi-Gang Cao	Beijing Jiaotong University, China
Hong-Wei Gao	Qingdao University, China
Wei-Bin Han	South China Normal University, China
Yi-Quan Li	Fuzhou University, China
Chao-Hui Wu	Fuzhou University, China
Genjiu Xu	Northwestern Polytechnical University, China
Xiao-Guang Yang	Academy of Mathematics and Systems Science, Chinese Academy of Sciences, China

Academic Program Committee

Chair

Deng-Feng Li	School of Economics and Management, Fuzhou University, China

Academic Program Committee Members

Youngsub Chun	Seoul National University, South Korea
Yukihiko Funaki	Waseda University, Japan
Toru Hokari	Keio University, Japan
Biung-Ghi Ju	Seoul National University, South Korea
Yuan Ju	York University, UK
Fuhito Kojima	Stanford University, USA
Jiu-Qiang Liu	Eastern Michigan University, USA

Shigeo Muto	Tokyo University of Science, Japan
Cheng-Zhong Qin	University of California, Santa Barbara, USA
Toyotaka Sakai	Keio University, Japan
Hao Sun	Northwestern Polytechnical University, China
Satoru Takahashi	National University of Singapore, Singapore
Robert Veszteg	Waseda University, Japan
Kang Xie	Sun Yat-Sen Business School, Guangzhou, China
Takehiko Yamato	Tokyo Institute of Technology, Japan
Xiao-Guang Yang	Academy of Mathematics and Systems Science, Chinese Academy of Sciences, China
Chun-Hsien Yeh	Academia Sinica, Taiwan, China
Qiang Zhang	Beijing Institute of Technology, China

Organizers

School of Economics and Management, Fuzhou University, China
Northwestern Polytechnical University, China

Sponsors

Social Science Research Management Department of Fuzhou University, China
The Innovation Team Leaded by Prof. Deng-Feng Li of School of Economics and Management, Fuzhou University, China

Contents

The Consensus Games for Consensus Economics Under the Framework of Blockchain in Fintech

Lan Di[1], Zhe Yang[2], and George Xianzhi Yuan[3,4,5,6,7](✉)

[1] School of Digital Media, Jiangnan University, Wuxi 214122, China
dilan@jiangnan.edu.cn
[2] School of Economics, Shanghai University of Finance and Economics, Shanghai 200433, China
zheyang211@163.com
[3] Business School, Chenngdu University, Chengdu 610106, China
george_yuan@yahoo.com
[4] School of Financial Technology, Shanghai Lixin University of Accounting and Finance, Shanghai 201209, China
[5] Center for Financial Engineering, Soochow University, Suzhou 215008, China
[6] Business School, Sun Yat-Sen University, Guangzhou 510275, China
[7] BBD Technology Co., Ltd. (BBD), No. 966, Tianfu Avenue, Chengdu 610093, China

Abstract. The goal of this paper is to introduce a new notion called "Consensus Game (CG)" with motivation from the mechanism design of blockchain economy under the consensus incentives from Bitcoin ecosystems in financial technology (Fintech), we then establish the general existence results for consensus equilibria of consensus games in terms of corresponding interpretation based on the viewpoint of Blockchain consensus in Fintech by applying the concept of hybrid solutions in game theory. As applications, our discussion in this paper for the illustration of some issues and problems on the stability of mining pool-games for miners by applying consensus games shows that the concept of consensus equilibria could be used as a fundamental tool for the study of consensus economics under the framework of Blockchain economy in Fintech.

Keywords: Hybrid solutions · Consensus equilibrium · Consensus game · Nakamoto consensus · Bitcoin ecosystem · Blockchain Protocol · Blockchain economy · Stability · Longest chain rules (LCR) · Chain Fork · Nonordered preferences · Mining economics · Minier dilemma · Multi-pools game · Fintech

This research is supported by National Natural Science Foundation of China (Nos. 11501349 and U181140002) in Part.

D. Li (Ed.): EAGT 2019, CCIS 1082, pp. 1–26, 2019.
https://doi.org/10.1007/978-981-15-0657-4_1

1 Introduction

In the game theory, Nash equilibrium follows the noncooperatitive idea, while the core is defined by considering the cooperative behavior of players. They concern on the noncooperative and cooperative idea respectively. Generally speaking, a cooperative solution concept (the α–core) was first introduced by Aumann [5]. Later, Scarf [44] proved an nonemptiness result for the α–core in a normal-form game with continuous and quasiconcave payoff functions. Inspired by Scarf [44], Kajii [29] provided a generalization of Scarf [44] to games with nonordered preferences, and Kajii's result and proof technique is also a modification and development of Border [11]. On the other hand, Florenzano [20] provided a new proof technique to obtain an existence theorem of the core in a coalition production economy with nonordered preferences. In [20], Florenzano defined the group preferences of each coalition and gave the proof by using Gale and Mas-Colell fixed point theorem. Following the method of [20], Lefebvre [32] provided a generalization to an economy with different information, and Martins-da-Rocha and Yannelis [33] extended Kajii's result to games on Hausdorff topological vector spaces. For more work on the α–core, we can refer to [1–4, 28, 38, 39, 45, 48, 52–54] and references wherein.

With motivation from Zhao [57] and Kajii [29], as discussed by Yang and Yuan [54] recently, our first goal is to establish consensus games without ordered preferences from the viewpoint of Blockchain in Fintech. In briefly, the consensus game considers whether there exists an acceptable (may or may not be "optimal") collaborating strategy which consists of a partial cooperative strategy and a partial noncooperative strategy under a given consensus rule in which some participants are based on cooperative, and the other part based on noncooperative game strategies to follow "Mining Longest Chain Rules (LCR)" (see also the original idea due to Nakamoto [36], the study by Nyumbayire [40], Biais [9] and reference wherein), while with or without occurring forks for blockchain acting as a platform (in supporting for different types business activities so-called digital economy). Thus, comparing with the traditional cooperative and noncooperative game, the consensus game is a natural extension for consensus economy, especially under the framework of Bitcoin ecosystem associated with consensus incentives in Bitcoin ecosystems (in terms of Nakamoto's consensus protocol as one example). Moreover, by following the study for the stability of the blockchain in supporting Bitcoin ecosystems under general consensus (due to Nakamoto [36]) in terms of mining economics, mining games and pool games extensively studied by Kroll et al. [31], Eyal and Sirer [19], Eyal [18], Bonneau et al. [10] (see also Carlsten et al. [12], Kiayias et al. [30], Sapirstein et al. [43], Biais et al. [9] and references wherein), plus the study on the existence of equilibria for blockchain disruption with or without occurring forks by Biais [9], smart contracts discussed by Cong and He [15], and move toward blockchain-based accounting and assurance given by Dai and Vasarhelyi [17], and following the idea of Zhao [57], it seems that the notion of consensus equilibria for consensus games with a partition of the set of players through the nonordered preferences mappings and related forms will be a useful tool for the study of consensus economics under the framework of Blockchain as a new kind of data structure in the practice.

Note that the classical results for the α–core concern on games with finitely many players. By considering a strong blocking concept, Weber [50] first proved an nonemptiness result for a core of a nontransferable cooperative game with infinitely many players. Inspired by Weber [50], Askoura [1] proved the existence of the weak core for games with a continuum of players. Later, Askoura [4] improved the result by considering the equi-usc condition of payoffs. Moreover, the work on the weak α–core was studied by Yang [52] and Yang [53]. Recently, Yang and Yuan [54] provided a generalization of Zhao [57] to games with nonordered preferences and proved the existence of weak hybrid solutions with infinitely many players.

In brief, we shall introduce a concept called "Consensus Game" (CG) with motivation from the mechanism design for the blockchain in financial technology under the consensus incentives introduced by Nakamoto [36] (see also Biais [9], Cong and He [15], Narayanan et al. [37], Nyumbayire [40] and related references wherein). Starting from results by Zhao [57] to Yang and Yuan [54] where a number of existence results have been established for a general game, while our paper mainly captures the consensus idea of blockchain consensus in Fintech, and the work of Yang and Yuan [54] plays an important role in modelling the Blockchain in Fintech, e.g., see Yuan et al. [58] and references wherein.

We like to share with readers that in this paper, we give a outline how the issue and problems on the stability of pool-games (e.g., the Bitcoin economy) can be formulated as applications of consensus games by using the concept of consensus equilibria, which could be used as a fundamental tool for the study of consensus economics under general framework of Blockchain economy in Fintech.

The rest of this paper is organized as follows. In Sect. 2, we recall the model and results from Zhao [57]. Section 3 recalls the main results from Yang and Yuan [54]. Section 4 illustrates how our notion of consensus games can be used easily to study the stability for Bitcoin ecosystems as applications of consensus games for two-pool games and multi-pools games, and Sect. 5 is the conclusion.

2 The Concept of Hybrid Solution in Game Theory

In this section, we recall some definitions and results from Zhao [57] (see also Yang and Yuan [54] and recent references wherein). Which were also recalled in Sect. 2 of [54]. For the sake of completeness for reading, we give their statements for results on Hybrid solution in game theory.

Let $N = \{1, \cdots, n_0\}$ be the set of agents and $p = \{N_1, \cdots, N_{k_0}\}$ be a partition of N, i.e.,

$$N_1 \bigcup \cdots \bigcup N_{k_0} = N, \; N_i \bigcap N_j = \emptyset, \forall i \neq j.$$

Denote by $\mathcal{N}$ the set of all nonempty subsets of N and $\mathcal{N}_r$ the set of nonempty subsets of N_r for each $r = 1, \cdots, k_0$.

A normal-form game with a partition can be defined by

$$G = (N, p, (X_i, u_i)_{i \in N}),$$

where X_i is the strategy set of player i, and $X = \prod_{i \in N} X_i$, $X_S = \prod_{i \in S} X_i$, $X_{-S} = \prod_{i \notin S} X_i, \forall S \in \mathcal{N}$; $u_i : X \longrightarrow R$ is the utility function of agent i. A strategy $x^* \in X$ is said to be a hybrid solution of G if for any $N_r \in p$ and any $S \in \mathcal{N}_r$, there exists no $y_S \in X_S$ such that

$$u_i(y_S, z_{N_r - S}, x^*_{-N_r}) > u_i(x^*_{N_r}, x^*_{-N_r}),\ \forall i \in S,\ \forall z_{N_r - S} \in X_{N_r - S}.$$

The following result is a stronger version of Theorem 2 in Zhao [57] (see also Theorem 2.1 of Yang and Yuan [54]).

Theorem 2.1. *Suppose that a normal-form game with a partition*

$$G = (N, p, (X_i, u_i)_{i \in N})$$

satisfies the following conditions:

(i) for each $i \in N$, X_i is a nonempty convex compact subset of R^{m_i};
(ii) for each $i \in N$, u_i is continuous and quasiconcave on X.

Then there exists at least a hybrid solution of G.

Furthermore, Zhao [57] defined a general cooperative game with a partition

$$G = \{G_r(x_{-N_r}) = (N_r, (X_S, u_S(\cdot, x_{-N_r}))_{S \in \mathcal{N}_r}) | r = 1, \ldots, k_0\},$$

where $u_S(\cdot, x_{-N_r}) : X_S \longrightarrow R^{|S|}$ is a vector-valued utility function of the coalition $S \in \mathcal{N}_r$ for any $x_{-N_r} \in X_{-N_r}$ and any $r = 1, \ldots, k_0$. A point $x^* \in X$ is a hybrid solution of G if for any $r = 1, \ldots, k_0$ and any $S \in N_r$, there exists no $y_S \in X_S$ such that

$$u_S(y_S, x^*_{-N_r}) > (u_N(x^*_{N_r}, x^*_{-N_r}))_S.$$

The following result is Theorem 3 of [57] (see also Theorem 2.2 of Yang and Yuan [54]).

Theorem 2.2 (Zhao [57]). *Suppose that a general cooperative game G with a partition p satisfies the following conditions:*

(i) for any $r = 1, \cdots, k_0$ and any $x_{-N_r} \in X_{-N_r}$, $G_r(x_{-N_r})$ is balanced;
(ii) for any $i \in N$, X_i is a nonempty convex compact subset of R^{m_i};
(iii) for any $N_r \in p$ and any $S \in \mathcal{N}_r$, u_S is continuous on $X_S \times X_{-N_r}$ and $u_S(\cdot, x_{-N_r})$ is quasiconcave on X_S for any $x_{-N_r} \in X_{-N_r}$.

Then there exists a hybrid solution of G at least.

3 The Concept of Consensus Games and Related Results

In this section, as applications of hybrid solutions, we shall introduce a new concept called "Consensus Game" (in short, "CG"), which is used in consensus economics to describe what kind of general consensus (through the realization of mechanism design) will achieve incentive compatibility to fight noncooperative behaviors and the coalition of participants (agents) under the platform of Blockchain in financial technology. Then we will discuss the existence of general consensus games' equilibria by using the concept of hybrid solutions. For the related reference on Blockchain and related Nakamoto consensus [36], please see Kroll et al. [31], Eyal and Sirer [19], Eyal [18], Bonnean et al. [10] (see also Carlsten et al. [12]), Kiayias et al. [30], Sapirstein et al. [43], Biais et al. [9], Nyumbayire [40], Narayanan [37] and related references wherein).

In the Fintech, in particular under the Nakamoto consensus protocol introduced in Year 2008, one key issue is to find a set of rules (for consensus) to encourage agents (miners from mining pools) to follow rules truthfully under the corresponding (consensus) protocol which may be formulated as preference mappings for abstract economy model (see Yannelis and Prabhakar [51], Yuan [55] and references wherein), thus it is very important to study the stability of Blockchain consensus in terms of equilibria for miners (from ming pools) to follow the so-called "Mining LCR" (also see the discussion in Sect. 4 below) while with or without occurring of forks for blockchain of Bitcoin ecosystems, the some other issues needed to be considered are possible collusive equilibria and their behavior related to smart contracts, or dynamic equilibria under blockchain disruption as initially discussed by Cong and He [15], and some other issues such as emerging blockchain-based accounting and assurance outlined by Dai and Varsarhelyi [17], discussed by Narayanan et al. [37] and so on.

Using the framework of the blockchain and associated consensus mechanism, the stability for Blockchain can be formulated as the question to find a strategy for all miners of pools (for Bitcoins) to follow up "LCR behaviors" respect to either noncooperative or cooperative behaviors (see also the discussion given in Sect. 4.3), which is exactly the notion for the concept of "hybrid solution" for games given by Zhao [57]), we thus come to have the following definition for a Consensus Game (in short, "CG"):

Given a consensus **G** (by consisting of a number of rules), let $N = \{1, 2, \cdots, n_0\}$ be the set of agents and $p = \{N_1, \cdots, N_{k_0}\}$ be a partition of N (as defined above). For each $i \in N$, the mapping $u_i : X \longrightarrow R$ is the payoff function of player i determined by the rules of the consensus **G**, we say that a normal form of consensus game (CG) is just the following form:

$$CG := (\mathbf{G}, N, p, (X_i, u_i)_{i \in N})$$

We say the consensus game CG has a consensus equilibrium if the corresponding formal form of the game $(N, p, (X_i, u_i)_{i \in N})$ has a bybrid solution.

We note that by using the quorum function, Zappala et al. [56] used the term (that is, consensus game) to measure agents' degree of supporting for the formulation of that coalition to study decision problem, which is different from our motivation above.

Throughout the rest part of this paper, when mentioning the consensus game (CG), we always assume it associated with the consensus **G** and omit it if no confusion. We now have the defining consensus equilibria for the consensus games with nonordered preferences.

A consensus game can be defined by

$$CG = (N, p, (X(t))_{t \in N}, P),$$

where $p = \{N_r | r \in R\}$ is a partition of N, $X(t)$ is the strategy space of player t, and $X = \prod_{t \in N} X(t), X(S) = \prod_{t \in S} X(t), X(-S) = \prod_{t \notin S} X(t), \forall S \in \mathcal{N}$, $P(t, \cdot) : X \rightrightarrows X$ is the preference mapping of player t. A point $x^* \in X$ is a consensus equilibrium of CG if for any $N_r \in p$ and any $S \in \mathcal{N}_r$, there exists no $y(S) \in X(S)$ such that

$$\{y(S)\} \times X(N_r - S) \times \{x^*(-N_r)\} \subset P(t, x^*), \ \forall t \in S.$$

We now recall results from Yang and Yuan [54]. The following result is the consensus game's version due to Theorem 3.1 of Yang and Yuan [54].

Theorem 3.1 (Yang and Yuan [54]**).** *Suppose that a consensus game*

$$CG = (N, p, (X(t))_{t \in N}, P)$$

satisfies the following conditions:

(i) N is a finite set;
(ii) for each $t \in N$, $X(t)$ is a nonempty convex compact subset of R^{m_t};
(iii) for each $t \in N$, $P(t, \cdot)$ is convex-valued with open graph in $X \times X$, and $x \notin P(t, x)$ for any $x \in X$.

Then there exists at least a consensus equilibrium of CG.

Yang and Yuan [54] next gave an infinite dimensional version of Theorem 3.1, see Theorem 3.2 of [54]. Here we state it by using concept of consensus games.

Theorem 3.2 (Yang and Yuan [54]**).** *Suppose that a consensus game*

$$CG = (N, p, (X(t))_{t \in N}, P)$$

satisfies the following conditions:

(i) N is a finite set;
(ii) for each $t \in N$, $X(t)$ is a nonempty convex compact subset of a Hausdorff topological vector space $E(t)$;
(iii) for each $t \in N$, $P(t, \cdot)$ is convex-valued with open graph in $X \times X$ and $x \notin P(t, x)$ for any $x \in X$.

Then there exists at least a consensus equilibrium of CG.

As an application of Theorem 3.2, we have the following corollary which is indeed an extension of Theorem 2.1 into topological vector spaces.

Corollary 3.1 *Suppose that a normal-form game with a partition*

$$G = (N, p, (X_i, u_i)_{i \in N})$$

satisfies the following conditions:

(i) for each $i \in N$, X_i is a nonempty convex compact subset of a Hausdorff topological vector space E_i;
(ii) for each $i \in N$, u_i is continuous and quasiconcave on X.

Then there exists at least a hybrid solution of G (thus the consensus equilibrium of consensus game G).

We next recall the result with infinitely many players from Yang and Yuan [54]. Let N be a topological space. We define the set Ω by

$$\Omega = \{(N_r, S) | S \subseteq N_r, N_r \in p\},$$

for a consensus game $CG = (N, p, (X(t))_{t \in N}, P)$.

A member (N_r, S) of Ω consensus-blocks a strategy $x \in X$ if there exists $y(S) \in X(S)$ such that

$$\{y(S)\} \times X(N_r - S) \times \{x(-N_r)\} \subset P(t, x), \ \forall t \in S.$$

A member (N_r, S) of Ω strongly consensus-blocks a strategy $x \in X$ if there exist $y(S) \in X(S)$ and an open set V in $N \times X \times X$ such that

$$S \times \{x\} \times \{y(S)\} \times X(N_r - S) \times \{x(-N_r)\} \subset V \subset clV \subset Graph(P).$$

A strategy $x^* \in X$ is a (weak) consensus equilibrium of CG if every member of Ω cannot (strongly) consensus-block x^*.

The proofs of the following result can be found in Lemmas 3.1–3.3 and Theorem 3.4 of Yang and Yuan [54], and we state it as an existence result for weak consensus equilibria of consensus games in a general form.

Theorem 3.3 (Yang and Yuan [54]). *Suppose that a consensus game $G = (N, p, (X(t))_{t \in N}, P)$ satisfies:*

(1) N is a nonempty compact Hausdorff topological space.
(2) For each $t \in N$, $X(t)$ is a nonempty convex compact subset of a Hausdorff topological vector space $E(t)$.
(3) The correspondence P is convex-valued with open graph in $N \times X \times X$ and $x \notin P(t, x)$ for all $(t, x) \in N \times X$.

Then there exists at least a weak consensus equilibrium.

In this section, the consensus games' results are mainly based on the existence results of theoretical models in game theory first established by Yang and Yuan [54], we omit their proof in details by saving spaces here (for proof details, we refer to Yang and Yuan [54]).

Next as applications, we will discuss the general stability problems of mining pool-games for miners under the framework of Blockchain consensus for Bitcoin economics through the illustration of the general mining pool games related miners for Bitcoins as examples to show that the concept of consensus equilibria could be used as a fundamental tool for the study of consensus economics in Fintech.

4 The Consensus Games for Bitcoin Ecosystems

In this section, we first discuss the general stability problems related study from a number of literatures for mining pool-games of miners for Bitcoins consensus principle (due to Nakamoto introduced in year 2008) under the framework of Blockchain consensus for Bitcoin economics. Then as applications of the existence of consensus equilibria (mainly Corollary 3.1) as an example, we will see how general mining pool games related miners for Bitcoins can be easily illustrated by the concept of consensus equilibria, which could be used as a fundamental tool for the study of consensus economics in Fintech.

Bitcoin is by far the most successful decentralized digital currency after the presentation of white paper by Nakamoto [36]. Its backbone is the blockchain protocol which attempts to keep a consisted list of transactions in a peer-to-peer network. The goal of blockchain protocol is to solve the real distributed problem of agreement, and has the potential to support innovation and applications which require distributed computing across a network, and proving new forms of assets such as "Digital Assets" and others associated business activities with the "Blockchain" (a kind of new data structure) which can also be interpreted acting as a "platform" in supporting digital assets' trading, financing, and many other kinds of business activities. All of these new forms of digital business, digital services with the complex relationship that exists between them under the environment of Bitcoin and blockchain are called "Bitcoin Ecosystems" in general.

Bitcoin implements its incentive systems with a data structure called the *Blockchain* as mentioned above by following "*Blockchain Protocol.*" The key idea of Blockchain Protocol is a serialization of all Bitcoin transactions. It is a single global ledger maintained by an open distributed system. Since anyone can join the open system and participate in maintaining the blockchain, Bitcoin uses a *proof of work* mechanism to deter attacks: participation requires exerting significant compute resources. A participant that proves she or he has exerted enough resources with a proof of work is allowed to take a step in the protocol by generating a block. Participants are compensated for their efforts with newly minted Bitcoins. The process of creating a block is called mining, and the participants, *miners*.

In order to win the reward, many miners try to generate blocks. The system automatically adjusts the *difficulty* of block generation, such that one block is added every 10 min to the blockchain. This means that each miner seldom generates a block. Although its revenue may be positive in expectation, a miner may have to wait for an extended period to create a block and earn the actual Bitcoins. Therefore, miners form *mining pools*, where all members mine concurrently and they share their revenue whenever one of them creates a block.

Pools are typically implemented as a *pool manager* and a cohort of miners. The pool manager joins the Bitcoin system as a single miner. Instead of generating proof of work, it outsources the work to the miners. In order to evaluate the miners efforts, the pool manager accepts partial proof of work and estimates each miner's *power* according to the rate with which it submits such partial proof of work. When a miner generates a full proof of work, it is sent to the pool manager which publishes this proof of work to the Bitcoin system. The pool manager thus receives the full revenue of the block and distributes it fairly according to its members' power. Many of the pools are open by allowing any miner to join them using a public Internet interface.

In fact Bitcoin's blockchain protocol provides two incentives for miners: "*Block rewards*" and "*ransaction fees*," which are key drivers for Bitcoin ecosystems. The former accounts for the vast majority of miner revenues at the beginning of the system, but it is expected to transition to the latter as the block rewards dwindle. There has been an implicit belief that whether miners are paid by block rewards or transaction fees does not affect the security of the block chain. But Carlsten et al. [12] (see also Kroll et al. [31], Eyal and Sirer [19], Eyal [18], Bonneau et al. [10] and a number of related references wherein) show that this is not the case, their key insight is that with only transaction fees, the variance of the block reward is very high due to the exponentially distributed block arrival time, and it becomes attractive to fork a "*wealthy*" block to "*steal*" the rewards therein. They show that this results in an equilibrium with undesirable properties for Bitcoin's security and performance, and even non-equilibria in some circumstances. Moreover, They also study selfish mining and show that it can be profitable for a miner with an arbitrarily low hash power share, who is arbitrarily poorly connected within the network, or working by themselves (i.e., miners' behavior in noncooperation game's way, or saying "noncooperative mining behavior"). Thus we need to consider the stability of Bitcoin ecosystems, in particular, for the miners' working behavior and strategy (also called "*mining for Bitcoin*") in different pools for the implementation of the most important parts due to Nakamoto's consensus being so-called the "*proof of work*" mechanism in Bitcoin economics.

4.1 The Stability of Bitcoin Ecosystems

We know that Bitcoin is the first widely popular cryptocurrency with a broad user base and a rich ecosystem, all hinging on the incentives in place to maintain the critical Bitcoin blockchain. For blockchain which acts as a platform (or saying, a new kind of data structures, or a tool) in supporting businesses under the

Bitcoin ecosystem, a natural process leads participants of such systems to form pools where members aggregate their power and share the rewards. Experience with Bitcoin shows that the largest pools are often open, allowing anyone to join. On the other hand, it has long been known that a member can sabotage an open pool by joining but never sharing proofs of work. The pool shares its revenue with the attacker, and each of its participants earns less.

Thus open pools are susceptible to the classical block withholding attack (e.g., see Rosenfeld [42]), where a miner sends only partial proof of work to the pool manager and discards full proof of work. Due to the partial proof of work sent to the pool by the miner, the miner is considered a regular pool member and the pool can estimate its power. Therefore, the attacker shares the revenue obtained by the other pool members, but does not contribute. It reduces the revenue of the other members, but also its own.

By thinking of another case is that a game where pools use some of their participants to infiltrate other pools and perform such an attack, one of the special cases is where either two pools or any number of identical pools play the game and the rest of the participants are uninvolved. In both of these cases, one natural question to ask is: does there exist an situation (equilibrium) that constitutes a tragedy of the commons where the participating pools attack one another and earn less than they would have if none had attacked?

Moreover, by following Bonneau et al. [10], we face two opposing viewpoints on Bitcoin in strawman form. The first is that "*Bitcoin works in practice, but not in theory.*" A second viewpoint is that "*Bitcoin's stability relies on an unknown combination of socioeconomic factors which is hopelessly intractable to model with sufficient precision, failing to yield a convincing argument for the system's soundness.*"

By putting above two opposing viewpoints on Bitcoin together, and incorporating Bitcoin's three main (technical) components: "*Transactions (including scripts),*" "*Consensus protocol,*" and "*Communication network*" as a whole, we do think it is critical to study the **Stability** (with more details below) for Bitcoin respect to its three main components in terms of complex ecosystem.

As shown from the comprehensive study on the stability discussed by Bonneau et al. [10], the *"stability of the consensus protocol"* should be one of the most important concepts respect to following five issues (see also Garay et al. [24], Kroll et al. [31], Miller and LaViola [35] and references wherein):

1 Eventual consensus: At any time, all compliant nodes will agree upon a prefix of what will become the eventual valid blockchain. We cannot require that the longest chain at any moment is entirely a prefix of the eventual blockchain, as blocks may be discarded (become stale) due to temporary forks.

2 Exponential convergence: The probability of a fork of depth n is $O(2^{-n})$. This gives users high confidence that a simple "k confirmations" rule will ensure their transactions are permanently included with high confidence.

3 Liveness: New blocks will continue to be added and valid transactions with appropriate fees will be included in the blockchain within a reasonable amount of time.

4 Correctness: All blocks in the longest chain will only include valid transactions.

5 Fairness: On expectation, a miner with a proportion α (the computing power) of the total computational power will mine a proportion $\propto \alpha$ of blocks (assuming they choose valid blocks).

If all of these properties hold we can say that the system is "**stable**," but it isn't clear that all are necessarily required as asked by Bonneau et al. [10]. However, Nakamoto [36] originally argued that Bitcoin will remain stable as long as all miners follow their own economic incentives (see also Nakamoto [36] again), a property called "*Incentive Compatibility.*" But the concept "*Incentive Compatibility*" has never been formally defined in the context of Bitcoin or cryptocurrencies; its prevalence as a term likely stems from its intuitive appeal and marketing value. We consider "*Compliant Miners*" whose strategies are the so-called "*default mining longest chain rules*" (LCR) (see also discussion by Biais et al. [9] and related references wherein). In game-theoretic terms, if universal compliance were shown to be a Nash equilibrium, this would imply incentive compatibility for Bitcoin as no miner would have any incentive to unilaterally change strategy. This would imply a notion of (weak) stability if other equilibria exist and strong stability if universal compliance were the sole equilibrium.

On the other hand non-compliant strategies dominate compliance, we must ask whether the resulting strategy equilibrium leads to stability for the consensus protocol, thus there are many issues and problems for which we are facing as recalled below in three categories based on the existing literatures (e.g., see mainly from Bonneau et al. [10]):

(1) Stability with bitcoin-denominated utility: We need to ask *if simple majority compliance may not ensure fairness?* by an interesting non-compliant mining strategy which is temporary block withholding as discussed by Bahackm [8], Eyal and Sirer [19], Garay et al. [24]; *if majority compliance is an equilibrium with perfect information* as shown by Kroll et al. [31]; *if majority compliance may imply convergence and consensus* as discussed by Miller and LaViola [35] and Garay et al. [24].

Secondly, one of the most important situations is that *with a majority miner, if stability is not guaranteed*: indeed, it is well known that a single non-compliant miner who controls a majority of computational power could undermine fairness by collecting all of the mining rewards simply by ignoring blocks found by others and building their own chain which by assumption would grow to become the longest chain. The majority miner could separately choose to undermine liveness by arbitrarily censoring transactions by refusing to include them and forking if they appear in any other block. Finally, the majority miner could undermine both convergence and eventual consensus by introducing arbitrarily long forks in the block chain, potentially to reverse and double-spend transactions for profit. All of these strategies would result in nominal profits, but since these behaviors are detectable, they may not be in a rational miner's long-term interest.

We like to mention that for the stability in terms of issue "*if mining longest chain rules*" (LCR) was also discussed by Biais et al. [9] through the Markov chain method under the situation with, or without mining a fork at the same time (i.e., the weak stability). Thus it is very important to discuss if it is possible among miners (in terms of either coordination (cooperation) or noncooperation behavior) for Bitcoin blockchain on mining LCR (while, with, or without occurring forking) as indeed a fork can also occur even when some miners adopt a new version of the mining software that is incompatible with the current version (if miners fail to coordinate on the same software, this triggers a fork).

Furthermore, in line with Nakamoto [36], it is said that the Bitcoin blockchain protocols are prone to multiple equilibria with forks due to the strategic complementarities of miner's actions, is it true? We would ask, *is the stability there if miners collude?*, and the question, *is stability there if mining rewards decline?*

(2) Stability with externally-denominated utility: Results in the bitcoin-denominated utility model do not provide convincing justification of Bitcoin's observed stability in practice as we may face the issues such as (a) **Liquidity limits**: Currently, exchanges which trade Bitcoin for external currencies typically have low liquidity. Thus, an attacker may obtain a large number of bitcoins but be unable to convert them all into external value, or can only do so at a greatly reduced exchange rate; (b) **Exchange rates in the face of attack**: Some non-compliant strategies, particularly those that would affect stability in a visible way, might undermine public confidence and hence weaken demand for bitcoins in the short run; (c) **Long-term stake in bitcoin-denominated mining rewards**: Most large miners have an additional interest in maintaining Bitcoin's exchange rate over time because they have significant capital tied up in non-liquid mining hard ware which will lose value if the exchange rate declines. If miners expect they will maintain their share of mining power far into the future with low marginal costs, then they may avoid strategies which earn them more bitcoins but decrease the expected value of their future mining rewards.

(3) Stability with incentives other than mining income: At least two strategies have been analyzed which may be advantageous for a miner whose utility is not purely derived from mining rewards, they are *Goldfinger attacks* (see also Kroll et al. [31]); *Feather-forking* proposed by Miller [34].

One of the key issues for the miners in the ming pool needs to consider is *what could go wrong* for the situation called "*Mining Gap*" which means if without a block reward immediately after a block is found if there is zero expected reward for mining but nonzero electricity cost, then it would be unprofitable for any miner to mine?

Indeed as discussed by Carlsteb et al. [12], we know that effects of a mining gap lead to miners mining for a smaller and smaller fraction of the time between the arrival of blocks (with the difficulty dropping to compensate). Clearly, this would have a negative impact for Bitcoin security, as the effective hash power in the network would drop, and it would become easier for a malicious miner to fork. Of course, turning a rig on and off every ten minutes may be practically infeasible. Nevertheless, this analysis illustrates that strategic miners might look for ways to deviate when the default protocol would have them wasting electricity to mine a near-valueless block.

The goal of this part is to establish an outline of the question on the stability of Bitcoin system on Blockchain can be formulated as the existence problem of consensus equilibria under the framework of consensus games with focus on the consensus associated with Bitcoin ecosystem, we give a brief recalling for the description of basic mining economics by following "**three types of consensuses**" below.

4.2 The Basic Mining Economics

Success of the Bitcoin economy requires that Bitcoin's distributed protocols operate and remain stable. In this section we consider the stability of these protocols, under the assumption that players behave according to their incentives. The success of Bitcoin relies on "**three types of consensuses**":

1 Consensus about Rules: Players must agree on criteria to determine which transactions are valid. Only valid transactions will be memorialized in the Bitcoin log, but this requires agreement on how to determine validity.

2 Consensus about State: Players must agree on which transactions have actually occurred, that is, they must agree on the history of the Bitcoin economy, so that there is a common understanding of who owns which coin at any given time.

3 Consensus that Bitcoins are Valuable: Players must agree that Bitcoins have value so that players will be willing to accept Bitcoins in payment.

Each of these forms of consensus depends mutually on the other two. For example, it is hard to agree on the history without agreeing on the rules, and it is hard to believe in the value of a Bitcoin if participants cannot even agree on who owns which Bitcoin.

Consensus about the rules is a social process. Participants must come to a common understanding of what is allowed, so that the rules can be encoded into the software that each participant uses. In Bitcoin, small groups and individuals can exert outsized power.

Consensus about state is a technological problem in distributed systems design. Each player can see part of the state and the players need to cooperate, in large numbers and across a potentially unreliable network, to achieve a consistent understanding of the global state. Technological consensus must be achieved despite the possibility that some players will deviate from the published rules. In the distributed systems literature, devious behavior ("**Byzantine failures**") can often be tolerated if a sufficient majority of players are honest and cooperate. However, in Bitcoin, we explicitly assume that players will behave according to their incentives (assuming cooperation despite incentives to the contrary would make the design much simpler, though unrealistic.)

Game-theoretic issues are very important for the correct execution of the blockchain protocol. This was realized at its inception when its creator, Nakamoto [36] analyzed incentives in a simple, albeit insufficient, model. Understanding these issues is essential for the survival of bitcoin and the development of the blockchain protocol. In practice it can help understand their strengths and vulnerabilities and, in economic and algorithmic theory, it can provide an excellent example for studying how rational ("*selfish miners*") players can play games in a distributed way and map out their possibilities and difficulties.

Distilling the essential game-theoretic properties of blockchain maintenance is far from trivial; some "*attacks*" and vulnerabilities have been proposed but there seems to exist no systematic way to discover them. In this work, we will study two models' stutaions of mining pool-games as applications of our consensus games below in which the miners (the nodes of the distributed network that run the protocol and are paid for it) play a complete-information (may or may not be "*stochastic*") games. Although the miners in the actual blockchain game do not have complete information, our games aim to capture two important questions that selfish miners ask (see also Kiayias et al. [30], Carlsten et al. [12], Badertscher etal. [7] and related references wherein):

(a) What to compute next (more precisely, which block to mine);
(b) When to release the results of computation (more precisely, when to release a mined block).

Ideally, the blockchain would be a simple chain of blocks implying precedence between the corresponding transactions, i.e., a serialization of valid transactions between the clients of the Bitcoin protocol. This would be the case if miners always started mining at the last announced block and propagated each block creation immediately to the network of the remaining miners. However, the selfish nature of the miners who try to receive the rewards of as many blocks as possible (or even the inherent delay of block propagation in the distributed network) can result in temporary forks in the blockchain. The protocol suggests to the miners to always start mining at the end of the branch which needed the largest amount of computational effort so far, i.e., the end of the longest fork. This strategy is called "*frontier*" and we will call the miners that follow it "*honest miners.*"

The reward structure of the protocol guarantees that the honest miners revenue is proportional to their computational power. However, understanding when it is profitable for the miners to deviate from the honest strategy is a central question and has attracted a lot of attention. The original assumption was that no miner has an incentive to deviate from the honest strategy if the majority of the miners are honest. However, this is not true as shown by Eyal and Sirer [19]. They gave a specific strategy which, when followed by a miner with computational power at least 33% of the total power, provides rewards strictly better than the honest strategy (assuming that every other miner is honest). This was extended computationally by Sapirstein et al. [43].

Bitcoin is the first widely popular cryptocurrency with a broad user base and a rich ecosystem, all hinging on the incentives in place to maintain the critical Bitcoin blockchain. But Eyal and Sirer [19] show that Bitcoin's mining protocol is not "*incentive-compatible*": which means there exist a "*selfish-miner*", a mining strategy that enables pools of colluding miners that adopt it to earn revenues in excess of their mining power. As a result, higher revenues can lead new miners to join a selfish miner pool, a dangerous dynamic that enables the selfish mining pool to grow towards a majority. The Bitcoin system would be much more robust if it were to adopt an automated mechanism that can thwart selfish miners. Eyal and Sirer [19] suggest a backwards-compatible modification to Bitcoin that ensures that pools smaller than $\frac{1}{4}$ of the total mining power cannot profitably engage selfish mining, and they also show that at least $\frac{2}{3}$ of the network needs to be honest to thwart selfish mining, which concludes that "*a simple majority is not enough*", thus we need to study in which way the existence of equilibria for mining economics in "Mining LCR", while, with or without occurring "*mining fork*" from miners of the mining-pools.

4.3 The Stability of Mining Pool Games Under the Framework by Concepts of Consensus Games

Recently, there are a number of informal and/or ad hoc attempts to address the security of Bitcoin, an exciting recent line of work has focused on devising a rigorous cryptographic analysis of the system [e.g., see Garay et al. [23], Garay et al. [26], Pass et al. [41], Badertscher et al. [6]). At a high level, these works start by describing an appropriate model of execution, and, within it, an abstraction of the original Bitcoin protocol of Nakamoto (2008) along with a specification of its security goals in terms of a set of intuitive desirable properties (see Garay [25], Garay [26], Pass [41]), or in terms of a functionality in a simulation-based composable framework (see Badertscher [6]). They then prove that the Bitcoin protocol meets the proposed specification under the assumption that the majority of the computing power invested in mining bitcoins is by devices which mine according to the Bitcoin protocol, i.e., honestly. This assumption of honest majority of computing power which had been a folklore within the Bitcoin community for years underlying the system's security is captured by considering the parties who are not mining honestly as controlled by a central adversary who coordinates them trying to disrupt the protocol's outcome.

Meanwhile, a number of works have focused on a rational analysis of the system (see Rosenfeld [42], Carlsten [12], Eyal and Sirer [19] and references wherein). In a nutshell, these works treat Bitcoin as a game between the (competing) rational miners, trying to maximize a set of utilities that are postulated as a natural incentive structure for the system. The goal of such an analysis is to investigate whether or not, or under which assumptions on the incentives and/or the level of collaboration of the parties, Bitcoin achieves a stable state, i.e., a game-theoretic equilibrium. However, despite several enlightening conclusions, more often than not the prediction of such analyses is rather pessimistic. Indeed, these results typically conclude that, unless assumptions on the amount of honest computing powersometimes even stronger than just majority-are made, the induced incentives result in plausibility of an attack to the Bitcoin mining protocol, which yields undesired outcomes such as forks on the blockchain, or a considerable slowdown.

To our knowledge, no fork or substantial slowdown that is attributed to rational attacks has been observed to date, and the Bitcoin network keeps performing according to its specification, even though mining pools would, in principle, be able to launch collaborative attacks given the power they control. In the game-theoretic setting, this mismatch between the predicted and observed behavior would be typically interpreted as an indication that the underlying assumptions about the utility of miners in existing analysis do not accurately capture the miners rationale. With motivation of Badertscher et al. [7], we concern the following two situations with focus on the so-called "*Consensus Economics*" (which means the ecosystems based on the framework of Bitcoin consensuses in general):

Q1 Is Bitcoin possibly broken under different kinds of attacks (or, saying differently, why does Bitcoin ecosystem work and why do majorities not collude to break it)?
Q2 Why do honest miners keep mining given the plausibility of such attacks?

Indeed we may interpret the "attackers" as miners playing noncooperative games by taking different kinds of attack strategies, and "honest miners" playing cooperative games by following the "default compliant mining rule" of Bitcoin consensus. By putting Q1 and Q2 together, the existence of the Bitcoin ecosystem is equivalent to the existence of (hybrid) equilibrium which is the so-called "consensus equilibrium" of the "consensus game" defined above in this paper.

Therefore the existence of consensus equilibrium for consensus games under the general framework of Bitcoin consensus means there always exists a group of people working on the "Longest Chain Rule" (LCR) which assures the Blockchain under the Bitcoin consensus is properly maintained (though some miners working on forks, other miners do not, e.g., see also Biais et al. [9] from a different way to address the issue in terms of Markov perfect equilibrium). Thus the study for the existence of consensus equilibrium for consensus games provide the fundamental base for consensus economics in general. In this way, we can study the stability of mining games for Bitcoin as applications of the general existence results established for consensus games above in this paper as shown below.

4.4 The Miner's Dilemma and General Pool Game

By using the concept of consensus games established in this paper, we will discuss Miner's Dilemma for Two Pools Game and the general existence of stability for Multi-Pools Games which was first discussed by Eyal [18] in a different way. As applications of our new concept called "*Consensus Games*", we wish our discussion for the illustration of "Miner's Dilemma and General Pool Game" on the stability of mining pool-games for miners by applying consensus games provide an example that the concept of consensus equilibria could be used as a fundamental tool for the study of consensus economics under the framework of Blockchain economy in Fintech, e.g., see Yuan et al. [50] for the work on the existence of conesnsus equilibria for data trading under the framework of Internet of Things (IoT) with Blockchain Ecosystems as applications of our new notion of Consensus Games.

An open distributed system can be secured by requiring participants to present proof of work and rewarding them for participation. The Bitcoin digital currency introduced this mechanism, which has been adopted by almost all contemporary digital currencies and related services. A natural process leads participants of such systems to form pools, where members aggregate their power and share the rewards. Experience with Bitcoin shows that the largest pools are often open, allowing anyone to join. It has long been known that a member can sabotage an open pool by seemingly joining it but never sharing proofs of work. The pool shares its revenue with the attacker, and so each of its participants earns less.

As discussed by Eyal [18], we define and analyze a game where pools use some of their participants to infiltrate other pools and perform such an attack. With any number of pools, no-pool-attacks is not a Nash equilibrium. We study the special cases where either two pools or any number of identical pools play the game and the rest of the participants are uninvolved. In both of these cases there exists an equilibrium that constitutes a tragedy of the commons where the participating pools attack one another and earn less than they would have if none had attacked.

For a two-pools game, the decision whether or not to attack is called the "*Miner's Dilemma*", an instance of the iterative prisoner's dilemma. The game is played daily by the active Bitcoin pools, which apparently choose not to attack. If this balance breaks, the revenue of open pools might diminish, making them unattractive to participants.

The General Model

By following Eyal [18], we assume the Bitcoin system is comprised of the Bitcoin network and nodes with unique IDs, and progresses in steps. A node i generates tasks which are associated with its ID i. Denote the number of pools with p, the total number of mining power in the system with m and the miners participating in pool i, where $1 \leq i, j \leq p$ with m_i, and $m = \cup_{i=1}^{p} m_i$, and $m_i \cap m_j = \emptyset$ for each $i \neq j$.

A node can work on a task for the duration of a step. The result of this work is a set of partial proofs of work and a set of full proofs of work. The number of proofs in each set has a Poisson distribution, partial proofs with a large mean and full proofs with a small mean. Nodes that work on tasks are called a miners, miners have identical power, and hence identical probabilities to generate proofs of work.

The Bitcoin network pays for full proofs of work. To acquire this payoff an entity publishes a task and its corresponding proof of work to the network. The payoff goes to the ID associated with task. The Bitcoin protocol normalizes revenue such that the average total revenue distributed in each step is a constant throughout the execution of the system. Any node can transact Bitcoins to another node by issuing a Bitcoin transaction. Nodes that generate tasks but outsource the work are called pools. Pools send tasks to miners over the network, the miners receive the tasks, perform the work, and send the partial and full proofs of work to the pool.

We follow the same assumption used by Eyal [18], apart from working on tasks, we assume that all local operations, payments, message sending, propagation, and receipt are instantaneous, and we also assume that the number of miners is large enough such that mining power can be split arbitrarily without resolution constraints. We now recall two definitions as follows.

Definition 1 (A solo miner). A solo miner is a node that generates its own tasks. In every step it generates a task, works on it for the duration of the step and if it finds a full proof of work, it publishes this proof of work to earn the payoff.

Definition 2 (Revenue density). The revenue density of a pool is the ratio between the average revenue a pool member earns and the average revenue it would have earned as a solo miner.

We note that for a solo miner, its revenue density, and that of a miner working with an unattacked pool are one. If a pool is attacked with block withholding, its revenue density decreases.

The Pool Block Withholding Attack

Just as a miner can perform block withholding on a pool j, a pool i can use some of its mining power to infiltrate a pool j and perform a block withholding attack on j. Denote the amount of such infiltrating mining power at step t by $x_{i,j}(t)$. Miners working for pool i, either mining honestly or used for infiltrating pool j, are loyal to pool i. At the end of a round, pool i aggregates its revenue from mining in the current round and from its infiltration in the previous round. It distributes the revenue evenly among all its loyal miners according to their partial proofs of work. The pool's miners are oblivious to their role and they operate as regular honest miners, working on tasks.

The Pool Game

In the pool game, pools try to optimize their infiltration rates of other pools to maximize their revenue. The overall number of miners and the number of miners loyal to each pool remain constant throughout the game.

The Revenue Density Analysis of the Pool Game

Recall that m_i is the number of miners loyal to pool i. and $x_{i,j}(t)$ is the number of miners used by pool i to infiltrate pool j at step t. The mining rate of pool i is therefore the number of its loyal miners minus the miners it uses for infiltration. This effective mining rate is divided by the total mining rate in the system, namely the number of all miners that do not engage in block withholding. Denote the direct mining rate R_i of pool i at step t by

$$R_i = \frac{m_i - \Sigma_{j=1}^{p} x_{i,j}}{m - \Sigma_{j=1}^{p} \Sigma_{k=1}^{p} x_{j,k}}.$$

The revenue density $r_i(t)$ of pool i at the end of step t is its revenue from direct mining together with its revenue from infiltrated pools, divided by the number of its loyal miners together with block-withholding infiltrators that attack it:

$$r_i(t) = \frac{R_i(t) + \Sigma_{j=1}^{p} x_{i,j}(t) r_j(t)}{m_i + \Sigma_{j=1}^{p} x_{j,i}(t)}.$$

Hereinafter we move to a static state analysis and omit the t argument in the expressions.

It is clear that if no pool engages in block withholding, for any $i \in m_i$ and $j \in m_j$, we have

$$x_{i,j} = 0 \text{ and we also have that } r_i = \frac{1}{m}$$

that is, each miner's revenue is proportional to its power, be it in a pool or working solo.

The Case for One Attacker

Now by considering a simplified game of two pools, 1 and 2, where pool 1 can infiltrate pool 2, but pool 2 cannot infiltrates pool 1. The $m - m_1 = m_2$ miners outside both pools mine solo (or with closed pools that do not attack and cannot be attacked). The dashed red arrow indicates that $x_{1,2}$ of pool 1s mining power infiltrates pool 2 with a block withholding attack.

Since Pool 2 does not engage in block withholding, all of its m_2 loyal miners work on its behalf. Pool 1, on the other hand does not employ $x_{1,2}$ of its loyal miners, and its direct mining power is only $m_1 - x_{1,2}$ The Bitcoin system normalizes these rates by the total number of miners that publish full proofs, namely all miners but $x_{1,2}$. The pools direct revenues are therefore

$$R_1 = \frac{m_1 - x_{1,2}}{m - x_{1,2}}$$

$$R_2 = \frac{m_2}{m - x_{1,2}}.$$

The revenue density r_2 of pool 2 is

$$r_2 = \frac{R_2}{m_2 + x_{1,2}}$$

and the revenue r_1 of loyal Pool 1 miner is

$$r_1 = \frac{R_1 + x_{1,2} r_2}{m_1}$$

thus we have that

$$r_1 = \frac{m_1(m_2 + x_{1,2}) - x_{1,2}^2}{m_1(m - x_{1,2})(m_2 + x_{1,2})}.$$

The Case for Two Pools Game

We proceed to analyze the case where two pools may attack each other and the other miners mine solo. Again we have pool 1 of size m_1 and pool 2 of size m_2; pool 1 controls its infiltration rate $x_{1,2}$ of pool 2, but now pool 2 also controls its infiltration rate $x_{2,1}$ of pool 1. Thus total mining power in the system is $m - x_{1,2} - x_{2,1}$. The direct revenues R_1 and R_2 of the pools from mining are their effective mining rates, without infiltrating mining power, divided by the total mining rate:

$$R_1 = \frac{m_1 - x_{1,2}}{m - x_{1,2} - x_{2,1}}$$

$$R_2 = \frac{m_1 - x_{2,1}}{m - x_{1,2} - x_{2,1}}.$$

Then we have the revenues r_1 and r_2 in terms of $x_{1,2}$ and $x_{2,1}$ by the following formula:

$$r_1(x_{1,2}, x_{2,1}) = \frac{m_2 R_1 + x_{1,2}(R_1 + R_2)}{m_1 m_2 + m_1 x_{1,2} + m_2 x_{2,1}}$$

$$r_2(x_{2,1}, x_{1,2}) = \frac{m_1 R_2 + x_{2,1}(R_1 + R_2)}{m_1 m_2 + m_1 x_{1,2} + m_2 x_{2,1}}.$$

Now we have the following general existence result for Two-Pools games as an application of Theorem 3.1 by applying consensus games, this result was first given by Eyal [18].

Theorem 4.1 (Eyal [18]). *For a two-pools game, an equilibrium exists where neither pool 1 nor pool 2 can improve its revenue by changing its infiltration rate.*

Proof. For $i = 1, 2$, by the definition of r_i, we can verify that it is concave respect to $x_{1,2}$ and $x_{2,1}$. Then by Corollary 3.1, it follows that there exists a consensus equilibrium point for Two Pools game. The proof is complete.

For the Two Pools game, we remak that like what said by Eyal [18], no-attack is not an equilibrium point for the Two Pools game as each pool can increase its revenue by choosing a strictly positive infiltration rate. Thus $x_{1,2} = x_{2,1} = 0$ is not a solution for a Two Pools game. Secondly, it is easy to see that a pool improves its revenue compared to the no-pool-attacks scenario only when it controls a strict majority of the total mining power. This fact is also confirmed by the numerical results given by Eyal [18], and the corresponding numerical results show that in extreme cases a pool does not attack its counterpart, which means at equilibrium a pool will refrain from attacking only if the other pool is larger than about 80% of the total mining power.

As a consequence of Theorem 4.1, we have the following result called "Miner's Dilemma of Two Pools Game" (see also discussion by Eyal [18]).

Theorem 4.2 (Miner's Dilemma of Two Pools Game). *For Two Pools game, it exists "Miner's Dilemma" which means the revenue density of each pool is determined by the decision of both pools whether to attack or not. The dominant strategy of each player is to attack, however the payoff of both would be larger if they both refrain from attacking.*

Proof. Indeed by the fact that for the Two Pools game and $i = 1, 2$, when $x_{1,2} = x_{2,1} = 0$, the revenue densities are $r_1 = r_2 = 1$, but this is not an equilibrium point as shown by Theorem 4.1 (or by a simply fact that each pool can increase its revenue r_i by choosing a strictly positive infiltration rate $x_{1,2}$ or $x_{2,1}$).

By following the same argument by Eyal [18], without loss of generality by considering pool 1, as we know that if pool 2 does not attack, pool 1 can increase its revenue above 1 (i.e., $r_1 > 1$) by attacking. If pool 2 does attack but pool 1 does not, we denote the revenue of pool 1 by $\hat{r_1}$. The exact value of $\hat{r_1}$ depends on the values of m_1 and m_2, but it is always smaller than one. As we have seen above, if pool 1 does choose to attack, its revenue increases, but does not surpass one. This completes our claim.

Miner's Dilemma for two pools tell us that the revenue density of each pool is determined by the decision of both pools whether to attack or not. The dominant strategy of each player is to attack, however the payoff of both would be larger if they both refrain from attacking. We also have that in a Two Pools (scenario) game, the revenue at the symmetric equilibrium is inferior to the no-one-attacks non-equilibrium strategy.

Next we can establish the general existence in terms of consensus games for the general case of multi-pools games (which was called "*q Indentical Pool Games*" by Eyal [18]).

The General Multi-Pools Games

Let there be q pools of (identical) size m_i (for $i.j = 1, 2, \cdots q$, $m = \cup_{i=1}^{q} m_i$) (and $m_i \cap m_j = \emptyset$ for each $i \neq j$) that engage in block withholding strategies against one another. Other miners neither attack nor are being attacked, this is the general multi-pools game.

For a general multi-pools game, we would expect that there exists a symmetric equilibrium. Now by considering without loss of generality, a step of pool 1. It controls its attack rates each of the other pools, and due to symmetry they are all the same. We denote by $x_{1,-1}$, the attack rate of pool 1 against any other pool. Each of the other pools can attack its peers as well. Due to symmetry, all attack rates by all attackers are identical. Denote by $x_{-1,*}$ the attack rate of any pool other than 1 against any other pool, including pool 1.

We denote by R_1 the direct revenue (from mining) of pool 1 and by R_{-1} the direct revenue of each of the other pools. Similarly we denote by r_1 and r_{-1} the revenue densities of pool 1 and other pools, respectively. Then by following the similar ide used abovem we have the following

$$R_1 = \frac{m_i - (q-1)x_{1,-1}}{m - (q-1)(q-1)x_{-1,*} - (q-1)x_{1,-1}}$$

$$R_{-1} = \frac{m_i - (q-1)x_{-1,*}}{m - (q-1)(q-1)x_{-1,*} - (q-1)x_{1,-1}}$$

and

$$r_1 = \frac{R_1 + (q-1)x_{1,-1}r_{-1}}{m_i + (q-1)x_{-1,1}}$$

$$r_{-1} = \frac{R_{-1} + (q-2)x_{-1,*}r_{-1} + x_{-1,*}r_1}{m_i + (q-2)x_{-1,*} + x_{1,-1}}.$$

We note that in the symmetric case for general multi-pool game, it follows that $r_1 = r_{-1}$, and indeed like the two-pool game scenario, the revenue at the symmetric equilibrium is inferior to the no-one-attacks non-equilibrium strategy.

Now we have the following general existence consensus games for multi-pools games below.

Theorem 4.3. (General Multi-Pools Games). *For a given general multi-pools game, its consensus equilibrium always exists.*

Proof. For a given multi-pools games (assuming q pools with total number (of miners) being $m = \cup_{i=1}^{q} m_i$, and $m_i \cap m_j = \emptyset$ for each $i \neq j$, while $i, j = 1, 2, \cdots, q$), then by the definition of r_i and its formula above, it follows that r_i is continuous (differentiable) and concave in $x_{1,-1}$. Thus by an application of Corollary 3.1, it follows that exists at least one consensus equilibrium for the general multi-pools games. This completes the proof.

Before we close this part, what we like to share with readers is that the concept of consensus games introduced in this paper is very useful for the study of Bitcoin ecosystem for its stability in terms of the existence for the consensus equilibria. By the fact that the success of the Bitcoin economy requires that Bitcoins distributed protocols operate and remain stable under the stability of these protocols associated with behaviors of players (miners) through pools relying on the incentives derived by the consensuses in general, thus we do think the theory of consensus games should provide the base for consensus Economics under the framework of the Blockchain ecosystems in Fintech (in particular for Bitcoin economics).

Finally, we like to mention that the stability of gap games formulated by Tsabary and Eyal [47] can also be studied under the framework of our consensus games in a natural way by using the partition of miners from different mining pools, and we plan to conduct this research in our next project soon.

5 The Concluding Remarks

Inspired by the hybrid solution concept of Zhao [57] and the mechanism design for the blockchain in financial technology, we introduce the model of the consensus game. Although Zappala et al. [56] used the term "consensus game" for the study of multiagent systems on autonomous agents, which is different from the outlined model and related concepts we introduced and discussed here in this paper. The key point of our model is to analyze the choice of strategies in which there exist cooperative and noncooperative behaviors at the same time and the preference of each agent is nonordered. Finally, we shall end this paper with the following remarks.

Our consensus game and consensus equilibrium can be regarded as an application of hybrid solutions introduced by Zhao [57]. The main model and result of consensus games have been given by Yang and Yuan [54]. In [54], the consensus equilibrium was called by hybrid solution since the viewpoint of [54] bases the game theory and the purpose of [54] is to study the existence of hybrid solutions with nonordered preferences and infinitely many players. In this paper, our purpose is to analyze the consensus economy under the framework of blockchain in Fintech. The result of Yang and Yuan [54] becomes a good technique to model the framework of blockchain in Fintech. Thus, in Sect. 3, we recall the work of [54] and call the concepts of [54] by consensus games and consensus equilibria.

Finally, we conclude that the existence results of consensus equilibria for consensus games defined in this paper are useful and should provide the base for the study of consensus economics under the framework of Blockchain in fintech as shown by the discussion above. Furthering the study on different situations related to smart contracts for different kinds of digital business activities under the Blackchain with associated consensus' incentives (called "blackchain economy") should be one of the most important things in era of big data. We also note that recently some issue and problems related to topics in fintech have been studied by a number of scholars, for example, the dynamic equilibria under blockchain disruption was initially discussed by Cong and He [15], topics surrounding blockchain-based accounting and assurance was outlined by Dai and Varsarhelyi [17], and other related areas of interest issues were discussed by Narayanan et al. [37]. Moreover a number of issues and problems in Finteh have been recently addressed by Goldstein et al. [27], Chiu and Koeppl [14]. Foley et al. [21], Fuster et al. [22], Tang [46], Vallée and Zeng [49], D'Acunto et al. [16], Zhu [59], Chen et al. [13] and references wherein. By using the new consensus games, Yuan et al. [50] recently conducted some work on the existence of consensus equilibria for data trading under the framework of Internet of Things (IoT) with Blockchain Ecosystems.

Finally, our thinks also go to Miss Susan Bin for her editing service which led to the present version of the paper.

References

1. Askoura, Y.: The weak-core of a game in normal form with a continuum of players. J. Math. Econ. **47**(1), 43–47 (2011)
2. Askoura, Y., Sbihi, M., Tikobaini, H.: The exante α−core for normal form games with uncertainty. J. Math. Econ. **49**(2), 157–162 (2013)
3. Askoura, Y.: An interim core for normal form games and exchange economies with incomplete information. J. Math. Econ. **58**, 38–45 (2015)
4. Askoura, Y.: On the core of normal form games with a continuum of players. Math. Soc. Sci. **89**, 32–42 (2017)
5. Aumann, R.J.: The core of a cooperative game without sidepayments. Trans. Am. Math. Soc. **98**, 539–552 (1961)
6. Badertscher, C., Maurer, U., Tschudi, D., Zikas, V.: Bitcoin as a transaction ledger: a composable treatment. In: Katz, J., Shacham, H. (eds.) CRYPTO 2017. LNCS, vol. 10401, pp. 324–356. Springer, Cham (2017). https://doi.org/10.1007/978-3-319-63688-7_11

7. Badertscher, C., Garay, J., Maurer, U., Tschudi, D., Zikas, V.: But why does it work? A rational protocol design treatment of bitcoin. In: Nielsen, J.B., Rijmen, V. (eds.) EUROCRYPT 2018. LNCS, vol. 10821, pp. 34–65. Springer, Cham (2018). https://doi.org/10.1007/978-3-319-78375-8_2
8. Bahackm, L.: Theoretical Bitcoin Attacks with less than Half of the Computational Power. Technical Report abs/1312.7013, CoRR (2013)
9. Biais, B., Bisire, C., Bouvard, M., Casamatta, C.: The blockchain folk theorem. Rev. Financ. Stud. **32**(5), 1662–1715 (2019)
10. Bonneau, J., Miller, A., Clark, J., Narayanan, A., Kroll, A., Felten, E.: Research Perspectives and Challenges for Bitcoin and Cryptocurrencies. In: Proceedings of the 36th IEEE Symposium on Security and Privacy, San Jose, California, USA, May 18–20, 2015
11. Border, K.C.: A core existence theorem for games without ordered preferences. Econometrica **52**(6), 1537–1542 (1984)
12. Carlsten, M., Kalodner, H., Weinberg, S.M., Narayanan, A.: On the instability of Bitcoin without the block reward. In: Proceedings of the 2016 ACM SIGSAC Conference on Computer and Communications Security, ACM, pp. 154C67, Vienna, Austria, October 24–28, 2016
13. Chen, M., Wu, Q., Yang, B.: How valuable is FinTech innovation? Revi. Financ. Stud. **32**(5), 2062–2106 (2019)
14. Chiu, J., Koeppl, T.: Blockchain-based settlement for asset trading. Rev. Financ. Stud. **32**(5), 1716–1753 (2019)
15. Cong, L.W., He, Z.: Blockchain disruption and smart contracts. Rev. Financ. Stud. **32**(5), 1754–1797 (2019)
16. D'Acunto, F., Prabhala, N., Rossi, A.G.: The promises and pitfalls of Robo-Advising. Rev. Financ. Stud. **32**(5), 1983–2020 (2019)
17. Dai, J., Vasarhelyi, M.A.: Toward blockchain-based accounting and assurance. J. Inf. Syst. **31**, 5–21 (2017)
18. Eyal, I.: The miners dilemma. In: Proceedings of the 36th IEEE Symposium on Security and Privacy, San Jose, California, USA, May 18–20, 2015
19. Eyal, I., Sirer, E.G.: Majority is not enough: Bitcoin mining is vulnerable. In: Christin, N., Safavi-Naini, R. (eds.) FC 2014. LNCS, vol. 8437, pp. 436–454. Springer, Heidelberg (2014). https://doi.org/10.1007/978-3-662-45472-5_28
20. Florenzano, M.: On the nonemptiness of the core of a coalitional production economy without ordered preferences. J. Math. Anal. Appl. **141**, 484–490 (1989)
21. Foley, S., Karlsen, J.R., Putnins, T.: Sex, Drugs, and Bitcoin: how much illegal activity is financed through Cryptocurrencies? Rev. Financ. Stud. **32**(5), 1798–1853 (2019)
22. Fuster, A., Plosser, M., Schnabl, P., Vickery, J.: The role of technology in mortgage lending. Rev. Financ. Stud. **32**(5), 1854–1899 (2019)
23. Garay, J.A., Katz, J., Tackmann, B., Zikas, V.: How fair is your protocol? A utility-based approach to protocol optimality. In: Georgiou, G., Spirakis, P.G. (eds.) The 34th ACM PODC, ACM. pp. 281–290, July 2015
24. Garay, J.A., Kiayias, A., Leonardos, N.: The Bitcoin Backbone Protocol: Analysis and Applications. Cryptology ePrint Archive, Report 2014/765 (2014)
25. Garay, J., Kiayias, A., Leonardos, N.: The bitcoin backbone protocol: analysis and applications. In: Oswald, E., Fischlin, M. (eds.) EUROCRYPT 2015. LNCS, vol. 9057, pp. 281–310. Springer, Heidelberg (2015). https://doi.org/10.1007/978-3-662-46803-6_10

26. Garay, J., Kiayias, A., Leonardos, N.: The bitcoin backbone protocol with chains of variable difficulty. In: Katz, J., Shacham, H. (eds.) CRYPTO 2017. LNCS, vol. 10401, pp. 291–323. Springer, Cham (2017). https://doi.org/10.1007/978-3-319-63688-7_10
27. Goldstein, I., Jiang, W., Karolyi, G.: To FinTech and beyond. Rev. Financ. Stud. **32**(5), 1647–1661 (2019)
28. Ichiishi, T.: A social coalitional equilibrium existence lemma. Econometrica **49**, 369–377 (1981)
29. Kajii, A.: A generalization of Scarf's theorem: an $\alpha-$core existence theorem without transitivity or completeness. J. Econ. Theory **56**, 194–205 (1992)
30. Kiayias, A., Koutsoupias, E, Kyropoulou, M, Tselekounis, Y.: Blockchain mining games. In: 2016 ACM Conference on Economics and Computation, Maastricht, The Netherlands, 24–28 July 2016
31. Kroll, J., Davey, I., Felten, E.: The economics of Bitcoin mining, or Bitcoin in the presence of adversaries. In: Proceedings of The Twelfth Workshop on the Economics of Information Security (WEIS 2013), Georgetown University, Washington DC, USA, 11–12 June 2013
32. Lefebvre, I.: An alternative proof of the nonemptiness of the private core. Econ. Theor. **18**(2), 275–291 (2001)
33. Martins-da-Rocha, V.F., Yannelis, N.: Nonemptiness of the alpha core. Working paper. Manchester School of Social Sciences, University of Manchester (2011)
34. Miller, A.: Feather-forks: enforcing a blacklist with sub-50% hash power. bitcointalk.org, October 2013
35. Miller, A., LaViola Jr., J.J.: Anonymous byzantine consensus from moderately-hard puzzles: a model for Bitcoin (2014)
36. Nakamoto, S.: Bitcoin: a peer-to-peer electronic cash system (2008). http://bitcoin.org/bitcoin.pdf
37. Narayanan, A., Bonneau, J., Felten, E., Miller, A., Goldfeder, S.: Bitcoin and Cryptocurrency Technologies: A Comprehensive Introduction Hardcover. Princeton University Press, Princeton (2016)
38. Noguchi, M.: Cooperative equilibria of finite games with incomplete information. J. Math. Econ. **55**, 4–10 (2018)
39. Noguchi, M.: Alpha cores of games with nonatomic asymmetric information. J. Math. Econ. **75**, 1–12 (2018)
40. Nyumbayire, C.: The Nakamoto Consensus. https://www.interlogica.it/en/insight-en/nakamoto-consensus. Insight, Interlogica, February 2017
41. Pass, R., Seeman, L., Shelat, A.: Analysis of the blockchain protocol in asynchronous networks. In: Coron, J.-S., Nielsen, J.B. (eds.) EUROCRYPT 2017. LNCS, vol. 10211, pp. 643–673. Springer, Cham (2017). https://doi.org/10.1007/978-3-319-56614-6_22
42. Rosenfeld, M.: Analysis of Bitcoin pooled mining reward systems. arXiv preprint arXiv:1112.4980 (2011)
43. Sapirshtein, A., Sompolinsky, Y., Zohar, A.: Optimal selfish mining strategies in Bitcoin. In: Grossklags, J., Preneel, B. (eds.) FC 2016. LNCS, vol. 9603, pp. 515–532. Springer, Heidelberg (2017). https://doi.org/10.1007/978-3-662-54970-4_30
44. Scarf, H.E.: On the existence of a cooperative solution for a general class of n-person games. J. Econ. Theor. **3**, 169–181 (1971)
45. Shafer, W., Sonnenschein, H.: Equilibrium in abstract economies without ordered preferences. J. Math. Econ. **2**, 345–348 (1975)
46. Tang, H.: Peer-to-Peer lenders versus banks: substitutes or complements? Rev. Financ. Stud. **32**(5), 1900–1938 (2019)

47. Tsabary, I., Eyal, I.: The gap game. In: Proceedings of the 2018 ACM SIGSAC Conference on Computer and Communications Security (CCS 18) (2018)
48. Uyanik, M.: On the nonemptiness of the $\alpha-$core of discontinuous games: transferable and nontransferable utilities. J. Econ. Theor. **158**, 213–231 (2015)
49. Vallée, B., Zeng, Y.: Marketplace lending: a new banking paradigm? Rev. Financ. Stud. **32**(5), 1939–1982 (2019)
50. Weber, S.: Some results on the weak core of a non-side-payment game with infinitely many players. J. Math. Econ. **8**, 101–111 (1981)
51. Yannelis, N.C., Prabhakar, N.D.: Existence of maximal elements and equilibria in linear topological spaces. J. Math. Econ. **12**, 233–245 (1983)
52. Yang, Z.: Some infinite-player generalizations of Scarf's theorem: finite-coalition $\alpha-$cores and weak $\alpha-$cores. J. Math. Econ. **73**, 81–85 (2017)
53. Yang, Z.: Some generalizations of Kajii's theorem to games with infinitely many players. J. Math. Econ. **76**, 131–135 (2018)
54. Yang, Z., Yuan, X.Z.: Some generalizations of Zhao's theorem: bybrid solutions and weak hybrid solutions for games with nonordered preferences. J. Math. Econ. **84**, 94–100 (2019)
55. Yuan, X.Z.: The study of equilibria for abstract economies in topological vector spaces-a unified approach. Nonlinear Anal. **37**, 409–430 (1999)
56. Zappala, J., Alechina, N., Logan, B.: Consensus games. In: Proceedings of the 11th International Conference on Autonomous Agents and Multiagent Systems (AAMAS 2012), pp. 1309–1310 (2012)
57. Zhao, J.: The hybrid solutions of an N-person game. Games Econ. Behav. **4**, 145–160 (1992)
58. Yuan, X.Z, Di, L., Zeng, T.: The Existence of consensus games for data trading under the framework of Internet of Things (IoT) with blockchain ecosystems. Working Paper. School of FinTech, Shanghai Lixin University of Accounting and Finance (2019)
59. Zhu, C.: Big data as a governance mechanism. Rev. Financ. Stud. **32**(5), 2021–2061 (2019)

Characterizations of the Position Value for Hypergraph Communication Situations

Guang Zhang[1], Erfang Shan[2(✉)], and Shaojian Qu[1]

[1] Business School, University of Shanghai for Science and Technology, Shanghai 200093, People's Republic of China
[2] School of Management, Shanghai University, Shanghai 200444, People's Republic of China
efshan@shu.edu.cn

Abstract. We characterize the position value for arbitrary hypergraph communication situations. The position value is first presented by the Shapley value of the uniform hyperlink game or the k-augmented uniform hyperlink game, which are obtained from a given hypergraph communication situation. These results generalize the non-axiomatic characterization of the position value from communication situations in Kongo (2010) (Int J Game Theory (2010) 39: 669–675) to hypergraph communication situations. Based on the non-axiomatic characterizations, we further provide an axiomatic characterization of the position value for arbitrary hypergraph communication situations by employing component efficiency and a new property, named partial balanced conference contributions. Partial balanced conference contributions is developed from balanced link contributions in Slikker (2005) (Int J Game Theory (2005) 33: 505–514).

Keywords: Hypergraph communication situation · Position value · Characterization

1 Introduction

The study of TU-games with limited cooperation presented by means of a communication graph was initiated by Myerson (1977), and an allocation rule for such games, the so-called Myerson value, was also introduced simultaneously. Later on, various studies in this direction were done in the past forty years, such as Meessen (1988), Herings et al. (2008), van den Brink et al. (2011), van den Brink et al. (2012), Béal et al. (2012) and Shan et al. (2016). Among them, the allocation rule, named position value (Meessen 1988), is also widely studied for graph communication situations. Borm et al. (1992) provided an axiomatic

This research was supported in part by the National Nature Science Foundation of China (grant number 11971298, 71901145).

D. Li (Ed.): EAGT 2019, CCIS 1082, pp. 27–42, 2019.
https://doi.org/10.1007/978-981-15-0657-4_2

characterization of the position value for graph communication situations in which the underlying graph is cycle-free. However, the characterization for arbitrary graph communication situations was not completed until Slikker (2005). On the other hand, van den Nouweland et al. (1992) extended the position value to hypergraph communication situations. They also gave an axiomatic characterization of the position value for cycle-free hypergraph communication situations. Algaba et al. (2000) extended the position value to union stable systems and characterized it on a subclass union stable systems. However, an axiomatic characterization of the position value for arbitrary hypergraph communication situations has not yet been found and remains an open problem. Additionally, the approach of non-axiomatic characterization for the position value, which deserves to be mentioned, was studied in Casajus (2007) and Kongo (2010), respectively. Casajus (2007) showed that the position value can be expressed by the Myerson value on a modified game, called the link agent form (LAF), if the original game is a graph communication situation, and called the hyperlink agent form (HAF), whenever the original game is a hypergraph communication situation. In Kongo (2010), the method which studies the relationships between two values is called non-axiomatic characterization, and a non-axiomatic characterization of the position value on the class of graph communication situations is presented by using the divided link game. However, we noticed that the approach in Kongo (2010) does not work directly for hypergraph communication situations.

The main goal of this paper is to study both non-axiomatic and axiomatic characterization of the position value for hypergraph communication situations. In order to complete the non-axiomatic characterization of the position value, we introduce two new games obtained from a hypergraph communication situation, called the uniform hyperlink game and the k-augmented uniform hyperlink game, respectively. It turns out that the position value for hypergraph communication situations can be represented by the Shapley value on the corresponding uniform hyperlink game or the corresponding k-augmented uniform hyperlink game. Based on the above non-axiomatic characterizations, an axiomatic characterization for arbitrary hypergraph communication situations is proposed by component efficiency and a new property, called partial balance conference contributions. Component efficiency states that for each component of the hypergraph the total payoff to its players equals the worth of that component. Partial balanced conference contributions is developed from balanced link contributions which is used to characterize the position value for graph communication situations in Slikker (2005). The partial balanced conference contributions here also deals with the payoff difference a player experiences if another player breaks one of his hyperlinks. The intrinsical difference between the two balanced properties is whether the payoff difference of a player experiences is totally or partially attributing to another player.

This article is organized as follows. Basic definitions and notation are given in Sect. 2. Section 3 shows the main results of this paper. In Subsect. 3.1, we first introduce the concepts of the uniform hyperlink game and the k-augment uniform hyperlink game, and then by using the two modified games, we obtain two non-axiomatic characterizations of the position value on the class of hypergraph communication situations. The main property, called partial balanced conference contributions, is presented in Subsect. 3.2. According to the previous results in Subsect. 3.1, an axiomatic characterization of the position value is completed on the class of hypergraph communication situations by employing component efficiency and partial balanced conference contributions. In Sect. 4, we conclude this paper with some remarks.

2 Basic Definitions and Notation

In this section, we recall some definitions and concepts related to TU-games, hypergraph communication situations, and their allocation rules.

A *cooperative game with transferable utility*, or simply a *TU-game*, is a pair (N, v) where $N = \{1, 2, \ldots, n\}$ is a finite set of $n \geq 2$ players and $v : 2^N \rightarrow \mathbb{R}$ is a *characteristic function* defined on the power set of N with $v(\emptyset) = 0$. For any $S \subseteq N$, S is called a *coalition* and the real number $v(S)$ is the *worth* of coalition S obtained. A *subgame* of v with a nonempty set $T \subseteq N$ is a game $v_T(S) = v(S)$, for all $S \subseteq T$. We denote by $|S|$ the cardinality of $S \subseteq N$. A game (N, v) is *zero-normalized* if for any $i \in N$, $v(\{i\}) = 0$. Throughout this paper, we consider only zero-normalized games.

Let $\Sigma(N)$ be the set of all permutations on N. For any permutation $\sigma \in \Sigma(N)$, the corresponding marginal vector $m^\sigma(N, v) \in \mathbb{R}^n$ assigns to every player $i \in N$ a payoff $m_i^\sigma(N, v) = v(\sigma^i \cup \{i\}) - v(\sigma^i)$, where $\sigma^i = \{j \in N \mid \sigma(j) < \sigma(i)\}$ is the set of players preceding i in the permutation σ. The best-known single-valued solution, the *Shapley value* introduced in Shapley (1953), assigns to every (N, v) the average of all marginal vectors. Formally, the Shapley value is given as

$$Sh_i(N, v) = \frac{1}{|\Sigma(N)|} \sum_{\sigma \in \Sigma(N)} m_i^\sigma(N, v), \quad \text{for all } i \in N. \tag{1}$$

An alternative description of the Shapley value can be expressed by the Harsanyi dividends. In Shapley (1953), the *unanimity game* (N, u_T) according to $T \subseteq N$ is the game defined by $u_T(S) = 1$ if $T \subseteq S$, and $u_T(S) = 0$ otherwise. Then each game (N, v) can be written as a unique linear combination of unanimity games, i.e., $v = \sum_{T \in 2^N \setminus \{\emptyset\}} \lambda_T(v) u_T$ where $\lambda_T(v) = \sum_{S \subseteq T: S \neq \emptyset} (-1)^{|T|-|S|} v(S)$ is the *Harsanyi dividend* of a nonempty coalition $T \subseteq N$ (see, Harsanyi 1959). Therefore, the alternative expression of the Shaple value is given as follows,

$$Sh_i(N, v) = \sum_{T \subseteq N: i \in T} \frac{\lambda_T(v)}{|T|}, \quad \text{for all } i \in N. \tag{2}$$

The communication possibilities for a TU-game can be described by a (communication) *hypergraph* (N, H), where H is a family of non-singleton subsets of N, i.e., $H \subseteq \{e \subseteq N \mid |e| \geq 2\}$. The elements of N are called the *nodes* or *vertices* of the hypergraph that represent players, and the elements of H called *hyperlinks* or *hyperedges* represent *conferences* in which all players in a hyperlink have to be present before communication can take place (see, van den Nouweland et al. 1992). A hypergraph (N, H) is called *r-uniform* if $|e| = r$ for all $e \in H$. Clearly, a graph (N, L), $L \subseteq \{e \subseteq N \mid |e| = 2\}$, is a 2-uniform hypergraph, and in this case these hyperlinks are called *links*. Therefore, hypergraphs are a natural generalization of graphs in which hyperlinks may contain more than 2 nodes.

Let H_i be the set of hyperlinks containing player i in a hypergraph (N, H), i.e. $H_i = \{e \in H \mid i \in e\}$. The *degree* of i is defined as $d(i) = |H_i|$. A node $i \in N$ is *incident* with a hyperlink $e \in H$ if $i \in e$. Two nodes i and j of N are *adjacent* in hypergraph (N, H) if there is an hyperlink e in H such that $i, j \in e$. Two nodes i and j are *connected* if there exists a sequence $i = i_0, i_1, \ldots, i_k = j$ of nodes in (N, H) in which i_{l-1} is adjacent to i_l for $l = 1, 2, \ldots, k$. A *connected hypergraph* is a hypergraph in which every pair of nodes are connected. For a given hypergraph (N, H), a (connected) *component* of (N, H) is the maximal set of nodes of N in which every pair of nodes are connected. Let N/H denote the set of components in (N, H) and $(N/H)_i$ be the component containing $i \in N$. For any $S \subseteq N$, let $(S, H(S))$ be the *subhypergraph* induced by S, where $H(S) = \{e \in H \mid e \subseteq S\}$. A hypergraph (N, H') is called a *partial hypergraph* of (N, H) if $H' \subseteq H$. The notation $S/H(S)$ (or shortly S/H) and N/H' are defined similarly.

A *hypergraph communication situation*, or simply a *hypergraph game*, is a triple (N, v, H), where (N, v) is a zero-normalized TU-game and H is the set of hyperlinks in the hypergraph (N, H). In particular, if (N, H) is a graph, the triple (N, v, H) is called a *graph communication situation*, or a *graph game*. Let HCS^N denote the class of all hypergraph communication situations with fixed player set N.

A mapping f is called a *value* or *allocation rule* if it assigns to every hypergraph game $(N, v, H) \in HCS^N$ a payoff vector $f(N, v, H) \in \mathbb{R}$. In this paper, we introduce two well-known values for hypergraph communication situations. The *Myerson value* μ (Myerson 1977, 1980; van den Nouweland et al. 1992) is defined by

$$\mu(N, v, H) = Sh(N, v^H),$$

where $v^H(S) = \sum_{T \in S/H} v(T)$ for any $S \subseteq N$. The game (N, v^H) is called the *point game* or *hypergraph-restricted game*.

An alternative value, the *position value* π (Meessen 1988; van den Nouweland et al. 1992), is given by

$$\pi_i(N, v, H) = \sum_{e \in H_i} \frac{1}{|e|} Sh_e(H, v^N), \text{ for any } i \in N,$$

where $v^N(H') = \sum_{T \in N/H'} v(T)$ for any $H' \subseteq H$. The game (H, v^N) is called the *conference game* or *hyperlink game*.

3 The Characterizations of the Position Value

In this section we provide two kinds of characterizations of the position value for hypergraph communication situations: non-axiomatic characterization and axiomatic characterization. As for the non-axiomatic characterization, the position value is presented by the Shapley value on the uniform hyperlink game or on the k-augmented uniform hyperlink game, both of each are obtained from a given hypergraph communication situation. While, the axiomatic one shows that the position value is the unique solution satisfying component efficiency and a new property, named partial balanced conference contributions.

3.1 Non-axiomatic Characterizations of the Position Value

In order to achieve the non-axiomatic characterization, we first introduce the *uniform hyperlink game* induced by a hypergraph communication situation. The definition follows the spirits of the divided link game in Kongo (2010) and the hyperlink agent form (HAF) in Casajus (2007).

Definition 1. For any $(N, v, H) \in HCS^N$ without isolated player, its *uniform hyperlink game* $(U(H), w)$ is defined as follows: Let $\eta(H)$ denote the least common multiple of the numbers in $\{|e| \mid e \in H\}$. Then set

$$U(H)(i,e) = \{(i,e,k) \mid k = 1, 2, \ldots, \eta(H) \cdot |e|^{-1}\}, \tag{3}$$

$$U(H)(i) = \bigcup_{e \in H_i} U(H)(i,e), \quad U(H)(e) = \bigcup_{i \in e} U(H)(i,e),$$

$$U(H) = \bigcup_{i \in N} U(H)(i) = \bigcup_{e \in H} U(H)(e), \text{ and} \tag{4}$$

$$w(S) = v^N(H[S]) = \sum_{R \in N/H[S]} v(R), \text{ for all } S \subseteq U(H), \tag{5}$$

where $H[S] = \{e \in H \mid U(H)(e) \subseteq S\}$.

From the definition, it is clear that the new player set $U(H)$ is obtained from a hypergraph by expanding each node to several nodes in every its incident hyperlink. Every set $U(H)(i,e)$ consists of precisely $\eta(H) \cdot |e|^{-1}$ players of $U(H)$ which are derived by expanding the player $i \in e$ in hyperlink $e \in H_i$. $U(H)(i)$ is the set of players obtained by expanding $i \in N$ in all hyperlinks in H_i and $U(H)(e)$ is the set of players generated by expanding the members of $e \in H$. Therefore, we have $|U(H)| = \eta(H) \cdot |H|$ and $|U(H)(e)| = \eta(H)$ for any $e \in H$.

In order to understand the new player set efficiently, we have the following example.

Example 1. Consider a hypergraph (N, H), *where* $N = \{1, \ldots, 6\}$ *and* $H = \{e_1, e_2, e_3, e_4\}$, *in which* $e_1 = \{1, 4\}, e_2 = \{2, 5\}, e_3 = \{3, 6\}, e_4 = \{4, 5, 6\}$. *According to the definition of* $U(H)$, *we have the following sets (also see Fig. 1):*

$U(H)(1, e_1) = \{(1, e_1, 1), (1, e_1, 2), (1, e_1, 3)\}$;
$U(H)(2, e_2) = \{(2, e_2, 1), (2, e_2, 2), (2, e_2, 3)\}$;
$U(H)(3, e_3) = \{(3, e_3, 1), (3, e_3, 2), (3, e_3, 3)\}$;
$U(H)(4, e_1) = \{(4, e_1, 1), (4, e_1, 2), (4, e_1, 3)\}$; $U(H)(4, e_4) = \{(4, e_4, 1), (4, e_4, 2)\}$;
$U(H)(5, e_2) = \{(5, e_2, 1), (5, e_2, 2), (5, e_2, 3)\}$; $U(H)(5, e_4) = \{(5, e_4, 1), (5, e_4, 2)\}$;
$U(H)(6, e_3) = \{(6, e_3, 1), (6, e_3, 2), (6, e_3, 3)\}$; $U(H)(6, e_4) = \{(6, e_4, 1), (6, e_4, 2)\}$;

$$U(H)(i) = U(H)(i, e_i), \text{ for } i = 1, 2, 3;$$
$$U(H)(i) = \{U(H)(i, e_{\{i-3\}}), U(H)(i, e_4)\}, \text{ for } i = 4, 5, 6;$$
$$U(H)(e_i) = \{U(H)(i, e_i), U(H)(i+3, e_i)\}, \text{ for } i = 1, 2, 3;$$
$$U(H)(e_4) = \{U(H)(i, e_4)\}, \text{ for } i = 4, 5, 6;$$
$$U(H) = \bigcup_{i=1}^{6} U(H)(i) = \bigcup_{j=1}^{4} U(H)(e_j).$$

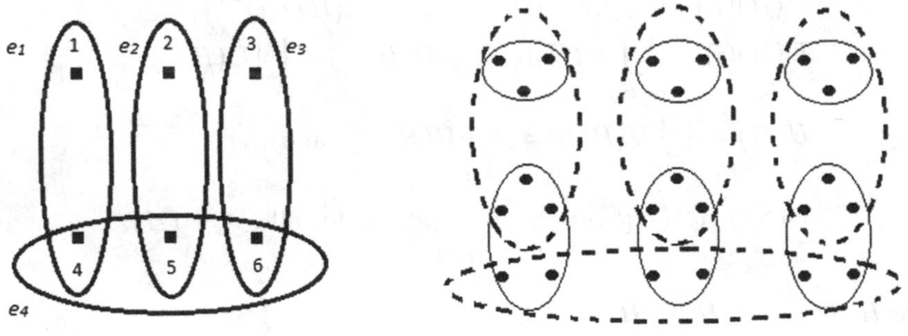

Fig. 1. The hypergraph (N, H) in Example 1 and its corresponding node sets.

Figure 1 shows that the subsets in which the nodes are encircled by solid lines are developed by the elements of N and indicated by $U(H)(i)$ for all $i \in N$, while the groups where the nodes are encircled by dotted lines are derived by the players in each hyperlink of H and represented by $U(H)(e)$ for all $e \in H$.

Remark. The definition of $U(H)$ is similar to the player set of HAF in Casajus (2007), but there are no hyperlinks or links on $U(H)$ which differs from HAF. In addition, the characteristic functions as in Definition 1 and as in HAF are different from each other as well. The characteristic function w follows the idea of the divided link game introduced in Kongo (2010). The uniform hyperlink game generalizes the divided link game from graph games to hypergraph games.

For a hypergraph game, we present the following non-axiomatic characterization of the position value in terms of the Shapley payoffs of the corresponding uniform hyperlink game.

Theorem 1. *For any hypergraph game $(N, v, H) \in HCS^N$ and $i \in N$, it holds that*

$$\pi_i(N, v, H) = \sum_{l \in U(H)(i)} Sh_l(U(H), w).$$

Proof. Let g be a mapping from $\Sigma(U(H))$ to $\Sigma(H)$: For any two hyperlinks $e_1, e_2 \in H$ and any permutation $\sigma \in \Sigma(U(H))$, $g(\sigma)(e_1) < g(\sigma)(e_2)$ if and only if $\max\{\sigma(l) \mid l \in U(H)(e_1)\} < \max\{\sigma(l) \mid l \in U(H)(e_2)\}$. Therefore, for any $l \in U(H)(e)$, if $\sigma(l) = \max\{\sigma(k) \mid k \in U(H)(e)\}$, then

$$m_l^{\sigma}(U(H), w) = m_e^{g(\sigma)}(H, v^N),$$

otherwise, we have $m_l^{\sigma}(U(H), w) = 0$. Hence,

$$\sum_{l \in U(H)(e)} m_l^{\sigma}(U(H), w) = m_e^{g(\sigma)}(H, v^N).$$

It is clear that $|\Sigma(U(H))| = (\eta(H) \cdot |H|)!$ and $|\Sigma(H)| = |H|!$. For each $\delta \in \Sigma(H)$, $\Sigma(U(H))$ has exactly $q = |\Sigma(U(H))|/|\Sigma(H)|$ permutations $\sigma_1, \sigma_2, \ldots, \sigma_q$ such that $g(\sigma_t) = \delta$ for $t = 1, 2, \ldots, q$. So, we have

$$\begin{aligned}
&\frac{1}{|\Sigma(U(H))|} \sum_{\sigma \in \Sigma(U(H))} \Big(\sum_{l \in U(H)(e)} m_l^{\sigma}(U(H), w) \Big) \\
&= \sum_{l \in U(H)(e)} \Big(\frac{1}{|\Sigma(U(H))|} \sum_{\sigma \in \Sigma(U(H))} m_l^{\sigma}(U(H), w) \Big) \\
&= \frac{1}{|\Sigma(H)|} \sum_{\delta \in \Sigma(H)} m_e^{\delta}(H, v^N). \qquad (6)
\end{aligned}$$

Therefore, by the definition of the Shapley value as in (1), we obtain

$$\sum_{l \in U(H)(e)} Sh_l(U(H), w) = Sh_e(H, v^N). \qquad (7)$$

Note that $|e| > 1$ for each $e \in H$, so there exist at least two players $l, l' \in U(H)(e)$ such that $w(S \cup \{l\}) = w(S) = w(S \cup \{l'\})$ for any $S \subseteq U(H) \setminus \{l, l'\}$. So l and l' are symmetric in $(U(H), w)$. From the symmetry of the Shapley value, it follows that $Sh_l(U(H), w) = Sh_{l'}(U(H), w)$ for any two players $l, l' \in U(H)(e)$. Therefore, according to (7), we have

$$Sh_l(U(H), w) = \frac{1}{\eta(H)} Sh_e(H, v^N), \text{ for any } l \in U(H)(e).$$

Note that, for each $e \in H$, it holds that $|U(H)(e)| = \eta(H)$. Therefore, it turns out that

$$\sum_{l \in U(H)(i,e)} Sh_l(U(H), w) = \frac{|U(H)(i,e)|}{\eta(H)} Sh_e(H, v^N) = \frac{1}{|e|} Sh_e(H, v^N).$$

The second equality holds by following (3) in Definition 1. Consequently,

$$\begin{aligned}\pi_i(N, v, H) &= \sum_{e \in H_i} \frac{1}{|e|} Sh_e(H, v^N) = \sum_{e \in H_i} \sum_{l \in U(H)(i,e)} Sh_l(U(H), w) \\ &= \sum_{l \in U(H)(i)} Sh_l(U(H), w),\end{aligned}$$

which completes the proof.

The following example illustrates Theorem 1.

Example 2. Consider a hypergraph communication situation (N, v, H), where $N = \{1, \ldots, 6\}$, v is the unanimity game $u_{\{1,2,3\}}$, and the underlying structure is the hypergraph in Example 1. Since the player set of the uniform hyperlink game is displayed in Example 1 as well, we focus our attention on the characteristic function w. According to (5) in Definition 1, we have

$$w(S) = \begin{cases} 1 & S = U(H), \\ 0 & \text{else}. \end{cases}$$

Hence, $Sh_l(U(H), w) = \frac{1}{24}$ for all $l \in U(H)$ and

$$\pi_1(N, v, H) = \sum_{l \in U(H)(1)} Sh_l(U(H), w) = \frac{1}{8} = \pi_2(N, v, H) = \pi_3(N, v, H),$$

$$\pi_4(N, v, H) = \sum_{l \in U(H)(4)} Sh_l(U(H), w) = \frac{3}{24} + \frac{2}{24} = \frac{5}{24} = \pi_5(N, v, H) = \pi_6(N, v, H).$$

So, this example shows that the position value of a hypergraph game can be expressed by the Shapley payoffs on the corresponding uniform hyperlink game.

By Definition 1 and the proof of Theorem 1, we note that $\eta(H)$ is the key point to guarantee (6) in the proof of Theorem 1. Clearly, if we consider an integral multiple of $\eta(H)$ instead of $\eta(H)$, (6) is still true. Based on this observation, we can construct another induced game from a given hypergraph game and describe the position value for the hypergraph game by the Shapley payoff on the induced games.

Definition 2. For any $(N, v, H) \in HCS^N$ and any positive integer $k \geq 1$, let $\eta(H)$ be the least common multiple of the numbers in $\{|e| \mid e \in H\}$ and $\rho(k) = k \cdot \eta(H)$. The *k-augmented uniform hyperlink game* $(U(H)^k, w^k)$ is defined as follows.

$$U(H)^k(i,e) = \{(i,e,t) \mid t = 1,\ldots,\rho(k)\cdot|e|^{-1}\}, \tag{8}$$

$$U(H)^k(i) = \bigcup_{e\in H_i} U(H)^k(i,e), \quad U(H)^k(e) = \bigcup_{i\in e} U(H)^k(i,e),$$

$$U(H)^k = \bigcup_{i\in N} U(H)^k(i) = \bigcup_{e\in H} U(H)^k(e), \text{ and} \tag{9}$$

$$w^k(S) = \sum_{R\in N/H[S]} v(R), \text{ for all } S \subseteq U(H)^k, \tag{10}$$

where $H[S] = \{e \in H \mid U(H)^k(e) \subseteq S\}$.

By the definition of the k-augmented uniform hyperlink game, we can obtain the following result which is more general than Theorem 1. Since the proof is same as in Theorem 1, we omit the details.

Theorem 2. *For any* $(N,v,H) \in HCS^N$ *and* $i \in N$, *it holds that*

$$\pi_i(N,v,H) = \sum_{l\in U(H)^k(i)} Sh_l(U(H)^k, w^k).$$

Clearly, the k-augmented uniform hyperlink game coincides with the uniform hyperlink game whenever $k = 1$, and Theorem 1 is a special case of Theorem 2. This result will be used to characterize the position value axiomatically for arbitrary hypergraph games in the next subsection.

3.2 The Axiomatic Characterization of the Position Value

In this subsection we provide an axiomatic characterization of the position value for arbitrary hypergraph communication situations.

Before introducing the properties used to characterize the position value, we first show an important lemma, which states that by deleting any hyperlink, the position value of the resulting hypergraph game can be expressed by the Shapley payoffs on a sub-uniform hyperlink game with respect to the grand coalition without any player derived from the deleted hyperlink.

Lemma 1. *For any hypergraph game* $(N,v,H) \in HCS^N$, $i \in N$, $e \in H$, *and* $l' \in U(H)(e)$, *we have*

$$\pi_i\big(N,v,H\setminus\{e\}\big) = \sum_{l\in(U(H)\setminus\{l'\})(i)} Sh_l\big(U(H)\setminus\{l'\}, w_{U(H)\setminus\{l'\}}\big).$$

Proof. To show the result, we distinguish two cases depending on whether the least common multiples $\eta(H)$ and $\eta(H\setminus\{e\})$ are same or not.

Case 1. $\eta(H\setminus\{e\}) = \eta(H)$. Then, by Theorem 1, we have

$$\pi_i(N,v,H\setminus\{e\}) = \sum_{l\in U(H\setminus\{e\})(i)} Sh_l(U(H\setminus\{e\}), w_{U(H\setminus\{e\})}). \tag{11}$$

Therefore, it is sufficient to show the following equality.

$$\sum_{l\in U(H\setminus\{e\})(i)} Sh_l(U(H\setminus\{e\}), w_{U(H\setminus\{e\})}) = \sum_{l\in(U(H)\setminus\{l'\})(i)} Sh_l(U(H)\setminus\{l'\}, w_{U(H)\setminus\{l'\}}). \tag{12}$$

By the definition of $U(H)$, we have $U(H\setminus\{e\}) = U(H)\setminus U(H)(e) \subseteq U(H)\setminus\{l'\}$ for any $e \in H$, where $l' \in U(H)(e)$. So, in $(U(H), w)$, for any $K \subseteq U(H)\setminus\{l'\}$, it follows that $w_{U(H)\setminus\{l'\}}(K) = w_{U(H\setminus\{e\})}(K \cap U(H\setminus\{e\})) = w(K)$. This means that the players in $\overline{U}(H, l') = (U(H)\setminus\{l'\})\setminus U(H\setminus\{e\})$ are null players of $w_{U(H)\setminus\{l'\}}$. Hence, we have that

$$Sh_l(U(H)\setminus\{l'\}, w_{U(H)\setminus\{l'\}}) = \begin{cases} Sh_l(U(H\setminus\{e\}), w_{U(H\setminus\{e\})}) & \text{if } l \in U(H\setminus\{e\}), \\ 0 & \text{if } l \in \overline{U}(H, l'). \end{cases} \tag{13}$$

This implies that (12) holds.

Case 2. $\eta(H\setminus\{e\}) \neq \eta(H)$. Since both $\eta(H)$ and $\eta(H\setminus\{e\})$ are the least common multiples of the numbers in $\{|e'| \mid e' \in H\}$ and $\{|e'| \mid e' \in H\setminus\{e\}\}$, respectively, it holds that $\eta(H) = |e|\cdot\eta(H\setminus\{e\})$. Therefore, by Theorem 2, we have

$$\pi_i(N, v, H\setminus\{e\}) = \sum_{l\in U(H\setminus\{e\})^k(i)} Sh_l(U(H\setminus\{e\})^k, w^k_{U(H\setminus\{e\})^k}). \tag{14}$$

where $k = |e|$. Thus, it is sufficient to show the following equality.

$$\sum_{l\in U(H\setminus\{e\})^k(i)} Sh_l(U(H\setminus\{e\})^k, w^k_{U(H\setminus\{e\})^k}) = \sum_{l\in(U(H)\setminus\{l'\})(i)} Sh_l(U(H)\setminus\{l'\}, w_{U(H)\setminus\{l'\}}).$$

Note that the remaining proof is similar to the proof in Case 1, so, above equality is true.

Summing up the two cases, it completes the proof of Lemma 1.

The next we show the expression of the position value which employs the Harsanyi dividends. For a given hypergraph communication situation $(N, v, H) \in HCS^N$, its uniform hyperlink game $(U(H), w)$ can be represented by a unique linear combination of unanimity games, i.e.,

$$w = \sum_{K\subseteq U(H)} \lambda_K(w) u_K. \tag{15}$$

By Theorem 1, the position value for (N, v, H) can be expressed in terms of the unanimity coefficients, i.e., Harsanyi dividends, of the associated uniform hyperlink game. Formally, for any $i \in N$, we have

$$\begin{aligned}\pi_i(N,v,H) &= \sum_{l\in U(H)(i)} Sh_l(U(H),w)\\ &= \sum_{l\in U(H)(i)} \sum_{K\subseteq U(H):l\in K} \frac{\lambda_K(w)}{|K|}\\ &= \sum_{K\subseteq U(H)} \lambda_K(w)\frac{|K_i|}{|K|}, \end{aligned} \tag{16}$$

where $K_i = K \cap U(H)(i)$ and the second equality follows from the alternative description of the Shapley value as in (2).

Let f be a value for hypergraph communication situations. We first recall a standard property for hypergraph games, called component efficiency, which was already used to characterize the Myerson value in Myerson (1977, 1980) and van den Nouweland et al. (1992). In addition, this property was also used to characterize the position value for graph games in Slikker (2005) and for cycle-free hypergraph games in van den Nouweland et al. (1992). It states that the total payoff of the players in a component is equal to the worth of this component.

Component efficiency: For any $(N, v, H) \in HCS^N$ and $T \in N/H$, it holds that

$$\sum_{i\in T} f_i(N,v,H) = v(T).$$

The second property is developed from balanced link contributions, which is proposed to characterize the position value for graph games in Slikker (2005). The property of balanced link contributions deals with the payoffs changes between two players. Formally, we have

Balanced link contributions: For any (N, v, L) and $i, j \in N$, it holds that

$$\sum_{e\in L_j} \big(f_i(N,v,L) - f_i(N,v,L\setminus\{e\})\big) = \sum_{e\in L_i} \big(f_j(N,v,L) - f_j(N,v,L\setminus\{e\})\big).$$

A natural extension of balanced link contributions from graph games to hypergraph games is the "balanced hyperlink contributions" or "balanced conference contributions", which can be seen in Shan et al. (2018) used to characterize the degree value for hypergraph games.

Balanced conference contributions: For any $(N, v, H) \in HCS^N$ and $i, j \in N$, it holds that

$$\sum_{e\in H_j} \big(f_i(N,v,L) - f_i(N,v,L\setminus\{e\})\big) = \sum_{e\in H_i} \big(f_j(N,v,L) - f_j(N,v,L\setminus\{e\})\big).$$

However, we note that the direct extension fails to characterize the position value axiomatically for hypergraph games and this point of view will be shown in Example 3. In order to characterize the position value for hypergraph games, we introduce a property called partial balanced conference contributions.

Partial balanced conference contributions also deals with the gains players contribute to each other. When a hyperlink related to a player is broken or built, the threat or contribution of another player received is not only depending on the first player, but also depending on those players whom adjacent to the first player according to the broken or built hyperlink. Formally, the property can be expressed as follows.

Partial balanced conference contributions: For any $(N, v, H) \in HCS^N$ and $i, j \in N$, it holds that

$$\sum_{e \in H_j} \frac{1}{|e|}(f_i(N, v, H) - f_i(N, v, H \setminus \{e\})) = \sum_{e \in H_i} \frac{1}{|e|}(f_j(N, v, H) - f_j(N, v, H \setminus \{e\})).$$

The partial balanced conference contributions states that the contribution or threat from a player towards another player equals the reverse contribution or threat, where the contribution or threat of a player towards another player is the sum of a portion payoff differences a player can inflict on another player by building or breaking one of his hyperlinks. In particular, if the underlying hypergraph is uniform, then the property coincides with the property of balanced conference contributions. However, this property is different from balanced hyperlink contributions in genenral.

Now, we show the position value satisfying the two properties: component efficiency and partial balanced conference contributions.

Lemma 2. *The position value for hypergraph communication situations satisfies component efficiency and partial balanced conference contributions.*

Proof. It has been verified that the position value satisfies component efficiency in van den Nouweland et al. (1992). We next show that the position value satisfies partial balanced conference contributions. Let $(N, v, H) \in HCS^N$ and $i, j \in N$ such that $i \neq j$. Then, we have

$$\begin{aligned}
&\sum_{e \in H_j} \frac{1}{|e|}(\pi_i(N, v, H) - \pi_i(N, v, H \setminus \{e\})) \\
&= \sum_{e \in H_j} \frac{|U(H)(j, e)|}{\eta(H)}(\pi_i(N, v, H) - \pi_i(N, v, H \setminus \{e\})) \\
&= \frac{1}{\eta(H)} \sum_{e \in H_j} \sum_{l \in U(H)(j,e)} \Big(\sum_{K \subseteq U(H)} \lambda_K(w) \frac{|K_i|}{|K|} - \sum_{K \subseteq U(H) \setminus \{l\}} \lambda_K(w) \frac{|K_i|}{|K|} \Big) \\
&= \frac{1}{\eta(H)} \sum_{l \in U(H)(j)} \sum_{K \subseteq U(H): l \in K} \lambda_K(w) \frac{|K_i|}{|K|} \\
&= \frac{1}{\eta(H)} \sum_{K \subseteq U(H)} |K_j| \lambda_K(w) \frac{|K_i|}{|K|}
\end{aligned}$$

$$= \frac{1}{\eta(H)} \sum_{K \subseteq U(H)} \lambda_K(w) \frac{|K_i| \cdot |K_j|}{|K|}$$

$$= \sum_{e \in H_i} \frac{1}{|e|} (\pi_j(N, v, H) - \pi_j(N, v, H \setminus \{e\})),$$

where the first equality follows from the definition of $U(H)(j,e)$ and $l \in U(H)(j,e)$, the second equality follows from (13) and Lemma 1 (note that $\lambda_K(w_{U(H)\setminus\{l\}}) = \lambda_K(w)$ for any $K \subseteq U(H) \setminus \{l\} \subseteq U(H)$), and the last equality follows from the symmetry of player i and j.

The following example illustrates that the position value satisfies partial balanced conference contributions and fails balanced conference contributions.

Example 3. Consider the hypergraph game as described in Example 2. The position payoffs for several hypergraph games are shown as follows.

$$\pi(N, v, A) = \begin{cases} (\frac{1}{8}, \frac{1}{8}, \frac{1}{8}, \frac{5}{24}, \frac{5}{24}, \frac{5}{24}) & A = H, \\ (0, 0, 0, 0, 0, 0) & A \subset H. \end{cases}$$

According to partial balanced conference contributions, the total contribution of player 6 to player 1 equals to $\frac{1}{2}(\frac{1}{8} - 0) + \frac{1}{3}(\frac{1}{8} - 0) = \frac{5}{48}$, *by breaking the hyperlink* e_3 *and* e_4, *respectively. The reverse contribution of player 1 to player 6 equals to* $\frac{1}{2}(\frac{5}{24} - 0) = \frac{5}{48}$ *as well by deleting the hyperlink* e_1. *Hence, the position value satisfies partial balanced conference contributions. However, the position value does not satisfies balanced conference contributions in this example. According to balanced conference contributions, the contribution of player 6 to player 1 equals to* $(\frac{1}{8} - 0) + (\frac{1}{8} - 0) = \frac{2}{4}$, *while, the reverse contribution of player 1 to player 6 equals* $\frac{5}{24} - 0 = \frac{5}{24}$.

We now present a characterization of the position value by component efficiency and partial balanced conference contributions. The proof is similar to the proof of Theorem 3.1 as in Slikker (2005), which characterizes the position value for arbitrary graph communication situations.

Theorem 3. *On the class of hypergraph communication situations, the position value is the unique allocation rule that satisfies component efficiency and partial balanced conference contributions.*

Proof. From Lemma 2, we know that the position value satisfies component efficiency and partial balanced conference contributions. Therefore, it only remains to prove the uniqueness, that is, there is a unique value satisfying the two properties. Suppose f is an allocation rule satisfies the two properties, we show that $f = \pi$. We proceed by induction on $|H|$. For $|H| = 0$, the assertion immediately follows from component efficiency. Next we may assume that f is determined and coincides with the position value π if $|H| \leq k-1$. We consider the case that $|H| = k$. For any component $C \in N/H$, let $C = \{1, 2, \ldots, c\}$. We obtain the following system of linearly independent equations by the two properties and the hypothesis,

$$\sum_{e\in H_2}\frac{1}{|e|}f_1(H)-\sum_{e\in H_1}\frac{1}{|e|}f_2(H)=\sum_{e\in H_2}\frac{1}{|e|}\pi_1(H\setminus\{e\})-\sum_{e\in H_1}\frac{1}{|e|}\pi_2(H\setminus\{e\}),$$

$$\cdots$$

$$\sum_{e\in H_c}\frac{1}{|e|}f_1(H)-\sum_{e\in H_1}\frac{1}{|e|}f_c(H)=\sum_{e\in H_c}\frac{1}{|e|}\pi_1(N,v,H\setminus\{e\})-\sum_{e\in H_1}\frac{1}{|e|}\pi_c(H\setminus\{e\}),$$

$$\sum_{i\in C}f(N,v,H)=v(C),$$

where write $f_i(H)$ and $f_i(H\setminus\{e\})$ instead of $f_i(N,v,H)$ and $f_i(N,v,H\setminus\{e\})$, respectively, for $i=1,2,\ldots,c$. One may easily verify that the above system has a unique solution. Since the position value satisfies component efficiency and partial balanced conference contributions, the position value is a solution of the above system. Consequently, we conclude that $f=\pi$ for any hypergraph communication situation with $|H|=k$.

4 Concluding Remarks

In this paper we provide two non-axiomatic characterizations and one axiomatic characterization of the position value for arbitrary hypergraph communication situations. Here the non-axiomatic characterization is in line with the works of Casajus (2007) and Kongo (2010). We noticed that Casajus (2007) expressed the position value for hypergraph communication situations in terms of the Myerson value by applying the hyperlink agent form (HAF). We now give a comparison between the expressions of the position value in Casajus (2007) and in our paper.

We first recall the definition of the hyperlink agent form $HAF(N,v,H)=(\bar{N},\bar{v},\bar{H})$, where the player set is

$$\bar{N}=\bigcup_{i\in N}\bar{N}(i),\quad \bar{N}(i)=\bigcup_{h\in H_i}\bar{N}(i,h),$$
$$\bar{N}(i,h)=\{(i,h,k)\,|\,k=1,2,\ldots,\eta(H)\cdot|h|^{-1}\},$$

the set of hyperlinks is

$$\bar{H}=\bar{H}^o\cup\bigcup_{i\in N}L^{\bar{N}(i)},\quad \bar{H}^o=\{\bar{h}\,|\,h\in H\},\quad \bar{h}=\bigcup_{i\in h}\bar{N}(i,h),$$

and the characteristic function is

$$\bar{v}(\bar{K})=v(N(\bar{K})),\quad N(\bar{K})=\{i\in N\,|\,\bar{N}(i)\cap\bar{K}\neq\emptyset\}.$$

According to the definitions of the HAF and the uniform hyperlink game, it is easy to check that the differences between the two induced games are in two aspects: the structures and the characteristic functions. However, somewhat surprisingly, we have the following relationships between them.

Corollary 1. *For any* $(N, v, H) \in HCS^N$, *it holds that* $Sh(U(H), w) = \mu(\bar{N}, \bar{v}, \bar{H})$, *where* $(U(H), w)$ *and* $(\bar{N}, \bar{v}, \bar{H})$ *are the uniform hyperlink game and the hyperlink agent form (HAF) of* (N, v, H), *respectively.*

Proof. By the definition of the player sets $U(H)$ and $\bar{N}$ and the Myerson value, it is sufficient to show that $w(S) = \bar{v}^{\bar{H}}(S)$ for any $S \subseteq U(H) = \bar{N}$.

In fact, for any $S \subseteq \bar{N}$, we have

$$\bar{v}^{\bar{H}}(S) = \sum_{C \in S/\bar{H}} \bar{v}(C) = \sum_{C \in S/\bar{H}} v(N(C)) = \sum_{C \in N(S)/H[S]} v(C)$$
$$= \sum_{C \in N/H[S]} v(C) = w(S),$$

where the third equality follows from the definition of $\bar{H}$ and $H[S]$, and the fourth equality holds due to the zero-normalized game.

Even though this corollary shows that the Shapley payoff of the uniform hyperlink game and the Myerson payoff of the HAF coincide with each other, the uniform hyperlink game is more concise than HAF. More importantly, the uniform hyperlink game can provide a powerful assistance in characterizing the position value for arbitrary hypergraph communication situations.

References

Algaba, E., Bilbao, J.-M., Borm, P., López, J.-J.: The position value for union stable systems. Math. Methods Oper. Res. **52**, 221–236 (2000)

Béal, S., Rémila, E., Solal, P.: Fairness and fairness for neighbors: the difference between the Myerson value and component-wise egalitarian solutions. Econ. Lett. **117**(1), 263–267 (2012)

Borm, P., Owen, G., Tijs, S.: On the position value for communication situations. SIAM J. Discrete Math. **5**, 305–320 (1992)

Casajus, A.: The position value is the Myerson value, in a sense. Int. J. Game Theory **36**, 47–55 (2007)

Harsanyi, J.-C.: A bargaining model for cooperative *n*-person games. In: Tucker, A.-W., Luce, R.-D. (eds.) Contributions to the Theory of Games IV, pp. 325–355. Princeton University Press, Princeton (1959)

Herings, P.-J.-J., van der Laan, G., Talman, A.-J.-J.: The average tree solution for cycle-free graph games. Games Econ. Behav. **62**(1), 77–92 (2008)

Kongo, T.: Difference between the position value and the Myerson value is due to the existence of coalition structures. Int. J. Game Theory **39**, 669–675 (2010)

Meessen, R.: Communication games. Master's thesis, Department of Mathematics. University of Nijmegen, the Netherlands (1988). (in Dutch)

Myerson, R.-B.: Graphs and cooperation in games. Math. Oper. Res. **2**, 225–229 (1977)

Myerson, R.-B.: Conference structures and fair allocation rules. Int. J. Game Theory **9**, 169–182 (1980)

Shan, E., Zhang, G., Dong, Y.: Component-wise proportional solutions for communication graph games. Math. Soc. Sci. **81**, 22–28 (2016)

Shan, E., Zhang, G., Shan, X.: The degree value for games with communication structure. Int. J. Game Theory **47**, 857–871 (2018)
Shapley, L.-S.: A value for n-person games. In: Kuhn, H., Tucker, A.-W. (eds.) Contributions to the Theory of Games II, pp. 307–317. Princeton, Princeton University Press (1953)
Slikker, M.: A characterization of the position value. Int. J. Game Theory **33**, 505–514 (2005)
van den Brink, R., Khmelnitskaya, A., van der Laan, G.: An efficient and fair solution for communication graph games. Econ. Lett. **117**(3), 786–789 (2012)
van den Brink, R., van der Laan, G., Pruzhansky, V.: Harsanyi power solutions for graph-restricted games. Int. J. Game Theory **40**, 87–110 (2011)
van den Nouweland, A., Borm, P., Tijs, S.: Allocation rules for hypergraph communication situations. Int. J. Game Theory **20**, 255–268 (1992)

A Class of Social-Shapley Values of Cooperative Games with Graph Structure

Hui Yang[1,2](✉), Hao Sun[1], and Genjiu Xu[1]

[1] Department of Applied Mathematics, Northwestern Polytechnical University, No. 127, Youyixilu, Xi'an, China
yangh816@mail.nwpu.edu.cn

[2] Department of Science, Xi'an University of Science and Technology, No. 58, Yanta Road, Xi'an, China

Abstract. This paper is devoted to a class of Social-Shapley values for cooperative games with graph structure. The Social-Shapley value compromises the utilitarianism of the Shapley value and the egalitarianism of the Solidarity value, in which the sociality is reflected by the Solidarity value. Through defining the corresponding properties in graph-restricted games, the paper axiomatically characterizes the Social-Shapley value when the coefficient is given exogenously. Moreover, we axiomatize the class of all possible Social-Shapley values in the graph-restricted games.

Keywords: Cooperative games · Graph structure · The Social-Shapley value · Axiomatization

1 Introduction

In the study of classical cooperative games, we often assume either that all players cooperate with each other, or else that the game is played noncooperatively. However, there are many intermediate possibilities between universal cooperation and no cooperation on account of the conflicts and competitions among participants. In this sense, only part of cooperation can be formed in the game.

Myerson [1] proposes the cooperative games with graph structure that the partial cooperation structure is depicted using an unordered graph. Given the participants set N, a graph on N is a set of unordered pairs of distinct members in N. We refer to these unordered pairs as links. The players may cooperate in a game by forming a series of bilateral agreements among themselves. These bilateral cooperative agreements can be represented by links between the agreeing players.

Figure 1 is taken from a wireless network and $N = \{1, 2, \ldots, 9\}$. The classical cooperation games could not exactly illustrate the cooperation relationships in this network while the graph theory offers a convenient method to describe the partial cooperation structure in the wireless network [2].

Naturally, the nodes in the network are denoted as the nodes in the graph and there exists a link between two nodes if they transmit the data directly.

D. Li (Ed.): EAGT 2019, CCIS 1082, pp. 43–56, 2019.
https://doi.org/10.1007/978-981-15-0657-4_3

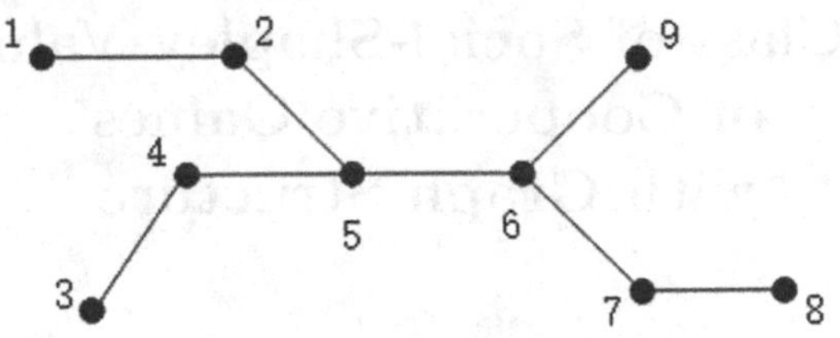

Fig. 1. An ad hoc wireless network

In the network every node can be endowed with a reputation value according to certain index, such as its connection degree or its packet forwarding performance for other nodes and so on [3]. If the reputation value of a node is smaller than others, which means its contribution is lower than others, other nodes may reject to forward the date toward this node afterwards. As a result, this node will be isolated gradually by others which will ultimately lead to the interruption of the whole wireless network. To avoid this situation, the node with lower reputation should be supported to some extent. This problem can be condensed to the payoff distribution of the cooperative game with graph structure.

The most well-known value in cooperative games is the Shapley value [4], which assigns to every player his expected marginal contribution assuming that all possible orders of the entrance of the players to the grand coalition occurs with equal probability. According to the Shapley value, the payoff of a dummy player only gets his individual value and a null player get nothing. Thus the vulnerable members could not get more through taking part in the cooperation. In this sense, the Shapley value is the embodiment of utilitarianism. In contrast, the Solidarity value [5] exchanges the marginal contributions with the average marginal contributions. The Solidarity value takes the social and psychological factors into account and benefit the vulnerable members in some extent, which embodies the idea of egalitarianism.

How to balance the relationship between egalitarianism and utilitarianism is a main issue in the economic allocation field. The convex combinations of solutions can efficiently reconcile the above two contradictory economic allocation thoughts. Joosten [6] introduces a class of convex combinations of the Shapley value and the equal division value, called as egalitarian Shapley values. Casajus and Huettner [7] characterize the class of the egalitarian Shapley values. Nowak and Radzik [8] introduced the convex combinations of the Shapley value and the Solidarity value under the circumstance of the classical cooperative games.

This paper names the convex combinations of the Shapley value and the Solidarity value the Social-Shapley value and generalizes the value to games with graph structure. The research can not only solve the partial cooperation structure between players but also balance the relationship between egalitarianism and utilitarianism. Besides, through defining some new axioms in games with graph structures, the paper axiomatically characterizes the Social-Shapley value. The first axiomatization concerns an axiom named α-counter contribution property in the case the coefficient is fixed exogenously. Then we substitute this property with rationality, the proportionality of Shapley value and the proportionality of Solidarity value to describe a class of Social-Shapley value in graph-restricted games.

The article is organized as follows. Section 2 presents the basic preliminaries about cooperative games and the Shapley value and the Solidarity value in graph-restricted games. Section 3 defines the Social-Shapley value in cooperative games with graph structure and axiomatically describes it. Based on this result, Sect. 4 axiomatically characterizes a class of the Social-Shapley values in graph-restricted games. Section 5 concludes the paper and gives the expected research of the relative value in graph-restricted games.

2 Preliminaries

A *cooperative game* with transferable utility (TU-game) describes a situation in which a finite set of players can obtain certain payoffs by cooperation. A TU-game is an ordered pair $\langle N, v\rangle$, where $N = \{1, 2, \ldots, n\}$ is a finite set of players with $|N| \geq 2$, called the *grand coalition*, and $v : 2^N \to \mathbb{R}$ is a *characteristic function* satisfying $v(\emptyset) = 0$. For any *coalition* $S \subseteq N$, $v(S)$ is the *worth* of coalition S, which is the utility the coalition S could receive when the players in S agree to cooperate with each other. The *cardinality* of coalition S is generically denoted by s, Particularly, n denotes the size of the player set N. The class of all TU games with the player set N is denoted by $\mathcal{G}^N$. Let $\langle N, v\rangle$, $\langle N, w\rangle \in \mathcal{G}^N$ and $c \in \mathbb{R}$, the games $\langle N, v+w\rangle$ and $\langle N, cv\rangle$ are defined by $(v+w)(S) = v(S)+w(S)$ and $(cv)(S) = cv(S)$, for all $S \subseteq N$.

A *value* φ is a function defined on $\mathcal{G}^N$ which associates any given game $\langle N, v\rangle$ a payoff vector $\varphi(N, v) = (\varphi_1(N, v), \varphi_2(N, v), \ldots, \varphi_n(N, v)) \in \mathbb{R}^N$. It uniquely determines a distribution of the payoffs available to the players in the game $\langle N, v\rangle$. The value also represents an assessment by i of his gains for participating in the game.

The well-known Shapley value [4] is defined as,

$$Sh_i(N, v) = \sum_{S \subseteq N, S \ni i} \frac{(n-s)!(s-1)!}{n!}[v(S) - v(S \backslash \{i\})], \quad i \in N.$$

The Shapley value just distributes to the dummy player his individual value, which could not illustrate the advantage of cooperation for the vulnerable participants.

Taking the social and psychological factors into account, the Solidarity value [5] assigns to every player in any game its expected gains when every marginal contribution is shared evenly among all players in the coalition by assuming that all possible orders of the entrance of the players to the grand coalition occur with equal probability. The *Solidarity value* is defined as

$$Sol_i(N, v) = \sum_{S \subseteq N, S \ni i} \frac{(n-s)!(s-1)!}{n!} A^v(S), \quad i \in N,$$

where $A^V(S) = \frac{1}{s}\sum_{k\in S}[v(S) - v(S\backslash\{k\})]$ is the *average marginal contribution* of coalition S.

If $v(S) - v(S\backslash\{i\}) > A^v(S)$, which means marginal contribution of player i is bigger than the average marginal contribution, player i will take out part of his marginal contribution to subsidize the vulnerable players in S, or else if $v(S) - v(S\backslash\{i\}) < A^v(S)$, player i will get support from others. Obviously, different from the Shapley value, the Solidarity value gives a positive payoff to the null player which benefits the vulnerable to a certain extent.

In order to compromising the severity of the Shapley value and the sociality of the Solidarity value, it seems reasonable to research the convex combinations of these two values. For any game $\langle N, v\rangle \in \mathcal{G}^N$, the convex combination of the Shapley value and the Solidarity value [8] is defined as:

$$\Phi_i^\lambda(N, v) = \lambda Sh_i(N, v) + (1-\lambda)Sol_i(N, v), \quad i \in N,$$

where $0 \leq \lambda \leq 1$. We can define this value as the *Social-Shapley value* in the reason that it combines the Shapley value and the social factor together. In this definition, the sociality is embodied by the Solidarity value. Nowak and Radzik [8] axiomatically describe this value under the circumstance of classical cooperative games.

Myerson [1] proposes the cooperative games with graph structure. Let $N = \{1, 2, \ldots, n\}$ be a finite set of players with $|N| \geq 2$. An *undirected graph* L on N without loops consists of a set of unordered pairs of distinct members of N, i.e., $L = \{\{i, j\}|i, j \in N, i \neq j\}$, where the *nodes* of the graph are identified with the players and $\{i, j\}$ correspond to the link between player i and j. We denote the set of all graphs on N by $g(N)$, then we present the basic terms of cooperative games with graph structure as follows [1].

1. For any $S \subseteq N$, $L \in g(N)$, i and j are *connected* in S by L if and only if there is a path in L which goes from i to j and stays within S.
2. S is a *connected coalition* if all the subgraphs in $S \subseteq N$ are connected. Given $L \in g(N)$, the set of all the connected coalitions in N is denoted by $C^L(N)$, similarly, the set of all the connected coalitions in $S \subseteq N$ is denoted by $C^L(S)$.
3. The largest connected coalition in $S \subseteq N$ is defined as the *connected component*, the set of connected components of S is a unique partition of S, which is denoted by S/L.

A *cooperative game with graph structure* is denoted by $\langle N, v, L\rangle$, where $N = \{1, 2, \ldots, n\}$ is the set of players, $L \in g(N)$ and $v : C^L(N) \to \mathbb{R}$ is a characteristic function. The class of all games with the player set N and graph structure L is denoted by $\mathcal{G}_N^L$. Without confusion, the game $\langle N, v, L\rangle \in \mathcal{G}_N^L$ is abbreviated as $\langle v, L\rangle$ for convenient. A value φ is defined on $\mathcal{G}_N^L$, $\varphi : \mathcal{G}_N^L \to \mathbb{R}^N$, which associates any game $\langle v, L\rangle$ a payoff vector $\varphi(v, L) = (\varphi_1(v, L), \varphi_2(v, L), \ldots, \varphi_n(v, L)) \in \mathbb{R}^N$.

In the graph-restricted games, the cooperative coalition can only be formed by the players in the connected coalition. Given the graph-restricted game $\langle v, L\rangle \in \mathcal{G}_N^L$, the corresponding graph-restricted game $\langle N, v^L\rangle \in \mathcal{G}^N$ can be defined as

$$v^L(S) = \sum_{C \in S/L} v(C).$$

Given the game with graph structure $\langle v, L\rangle \in \mathcal{G}_N^L$, for $i \in N$, $S \subsetneq N\backslash\{i\}$, i is a *dummy player* if $v^L(S \cup \{i\}) = v^L(S) + v^L(\{i\})$. Given any $\langle v, L\rangle \in \mathcal{G}_N^L$, the *Myerson value* is defined as

$$M_i(v, L) = Sh_i(N, v^L), \quad i \in N.$$

For any $\langle v, L\rangle \in \mathcal{G}_N^L$, the *Solidarity value in graph-restricted game* [9] is defined as

$$Sol_i(v, L) = \sum_{T \subseteq C, T \ni i} \frac{(c-t)!(t-1)!}{c!} A^{v^L}(T), \quad i \in N,$$

where the coalition C is the connected components including player i, $T \in C^L(N)$, $A^{v^L}(T) = \frac{1}{t}\sum_{k \in T}[v^L(T) - v^L(T\backslash\{k\})]$ is the average marginal contribution of players in T. Player i is *A-null player* if $A^{v^L}(T) = 0$ for $i \in T$, $T \in C^L(N)$.

For cooperative games with graph structure, a value φ is

- *Component Efficiency*: If $\sum_{i \in N} \varphi_i(v, L) = v^L(N)$ for all $\langle v, L\rangle \in \mathcal{G}_N^L$.
- *Symmetry*: If $\varphi_i(v, L) = \varphi_j(v, L)$, whenever i and j are symmetric players in $\langle v, L\rangle \in \mathcal{G}_N^L$.
- *Additivity*: If $\varphi_i(v + w, L) = \varphi_i(v, L) + \varphi_i(w, L)$ for all $\langle v, L\rangle$, $\langle w, L\rangle \in \mathcal{G}_N^L$.

3 The Social-Shapley Value in Graph-Restricted Games

Corresponding to the Social-Shapley value in classical cooperative games, this section researches the Social-Shapley value in graph-restricted games.

Definition 1. *Given a game $\langle v, L\rangle \in \mathcal{G}_N^L$, the Social-Shapley value in graph-restricted game is defined as*

$$\Phi_i^\lambda(v, L) = \lambda Sh_i(v, L) + (1-\lambda) Sol_i(v, L), \quad i \in N,$$

where the convex coefficient $\lambda \in [0, 1]$.

The smaller the coefficient λ is, the bigger sociality the $\Phi_i^\lambda(v, L)$ reflected. Plainly, we can define *the class of the Soical-Shapley values* as $\{\Phi_i^\lambda(v, L) | \lambda \in [0, 1]\}$.

The Shapley value distributes the payoff only based on a player's productivity measured by his own marginal contributions, while the Solidarity value reflects some social behavior of players in coalitions. The Social-Shapley value efficiently compromises these two values in graph-restricted games and can flexibly adjust the sociality of a payoff by modifying the value of coefficient λ through evaluating how friendly the players' relation is.

Given $\alpha \geq 0$ and the game $\langle v, L\rangle \in \mathcal{G}_N^L$, $S \subseteq N$, player i is *α-counter contribution player* if for any $i \in S$,

$$v^L(S) - v^L(S\backslash\{i\}) = -\alpha A^{v^L}(S). \tag{1}$$

- *α-counter contribution property*: Given any game $\langle v, L\rangle \in \mathcal{G}_N^L$, $\varphi_i(v, L) = 0$ if player i is α-counter contribution player.

Evidently, if the marginal contribution of player i to any coalition S is the negative of the average contribution of the members of S, up to a coefficient α, then he gets nothing from the game. This property can be used to characterize the Social-Shapley value in graph-restricted games.

Theorem 1. *Given $\alpha \geq 0$ and any game $\langle v, L\rangle \in \mathcal{G}_N^L$, the Social-Shapley value $\Phi^\lambda(v, L)$ is the unique value which satisfies component efficiency, additivity, symmetry and the α-counter contribution property, where*

$$\Phi_i^\lambda(v, L) = \lambda Sh_i(v, L) + (1-\lambda)Sol_i(v, L), \quad i \in N$$

and $\lambda = \frac{1}{1+\alpha}$.

In this section, the coefficient α is given arbitrarily. To prove Theorem 1, we introduce a special basis for the linear space $\mathcal{G}_N^L$ and two lemmas.

Let $\lambda = \frac{1}{1+\alpha}$. For any $T \in C^L(N)$, $T \neq \emptyset$, we defined the game $\langle w_T, L\rangle$ by

$$w_T^L(S) = \begin{cases} \frac{t!}{s!}\prod_{j=1}^{s-t}(\lambda t + j), & S \supseteq T; \\ 0, & \text{otherwise}, \end{cases}$$

specifically, assuming that $\prod_{j=1}^{0}(\lambda t + j) = 1$ and $w_T^L(T) = 1$. The games $\{\langle w_T, L\rangle | T \in C^L(N), T \neq \emptyset\}$ constitutes a basis of $\mathcal{G}_N^L$ (The proof is similar to Lemma 2.2 in paper [5]).

Lemma 1. *Let T be any non-empty coalition, $T \neq N$ and $S = T \cup D$, where $\emptyset \neq D \in C^L(N)$. Then for every player $i \in S\backslash T$, we have*

$$(1+\alpha)w_T^L(S) = w_T^L(S\backslash\{i\}) + \alpha\frac{1}{s}w_T^L(S\backslash\{k\}) \tag{2}$$

and each player $i \in N\backslash T$ satisfies the α-counter contribution property in the game $\langle w_T, L\rangle$.

Proof. Assuming that $d \geq 2$, where d is the cardinality of D, using the definition of $\langle w_T, L\rangle$, when $k \in T$, $w_T^L(S\backslash\{k\}) = 0$. As a result,

$$(1+\alpha)w_T^L(S) - w_T^L(S\backslash\{i\}) - \alpha\frac{1}{s}\sum_{k\in S} w_T^L(S\backslash\{k\})$$

$$= (1+\alpha)\frac{t!}{s!}\prod_{j=1}^{s-t}(\lambda t+j) - w_T^L(S\backslash\{i\}) - \alpha\frac{1}{s}\sum_{k\in D} w_T^L(S\backslash\{k\})$$

$$= (1+\alpha)\frac{t!}{(t+d)!}\prod_{j=1}^{d}(\lambda t+j) - \frac{t!}{(t+d-1)!}\prod_{j=1}^{d-1}(\lambda t+j)$$

$$-\alpha\frac{d}{t+d}\frac{t!}{(t+d-1)!}\prod_{j=1}^{d-1}(\lambda t+j)$$

$$= [(1+\alpha)(\lambda t+d) - (t+d) - \alpha d]\frac{t!}{(t+d)!}\prod_{j=1}^{d-1}(\lambda t+j)$$

$$= 0.$$

Thus, the lemma holds for $d \geq 2$. If $d = 1$, he above deduction exists because of the assumption of $\prod_{j=1}^{0}(\lambda t + j) = 1$.

Consider any player $i \in N\backslash T$, let S be an arbitrary coalition including player i. If T is not a subset of S, then both sides of Eq. (1) equal to zero with $\langle v, L\rangle = \langle w_T, L\rangle$, or else if T is a subset of S, then $i \in S\backslash T$, let $\langle v, L\rangle = \langle w_T, L\rangle$, Eqs. (1) and (2) are equivalent. Consequently, player i satisfies the α-counter contribution property. □

Lemma 2. *If Φ is a value on $\mathcal{G}_N^L$ satisfying component efficiency, additivity, symmetry and α-counter contribution property, then for any $T \in C^L(N)$, $T \neq \emptyset$ and $c \in \mathbb{R}$, we have*

$$\Phi_i(cw_T, L) = \begin{cases} \frac{c(t-1)!}{n!}\prod_{j=1}^{n-t}(\lambda t+j), & j \in T; \\ 0, & j \notin T, \end{cases}$$

specifically, $\Phi_i(cw_T, L) = \frac{c}{n}$ for each $i \in N$, when $T = N$.

Proof. For any coalition $T \neq \emptyset$ and $T \neq N$. If $c = 0$, the lemma holds according to the component efficiency and additivity. If $c \neq 0$, every player $i \in N\backslash T$ is a α-counter contribution player in the game $\langle cw_T, L\rangle$, therefore $\Phi_i(cw_T, L) = 0$ for each i by α-counter contribution property. Combining with the component efficiency and symmetry, the lemma holds. If $T = N$, the proof is trivial. Until now, the lemma has been proved. □

Based on Lemmas 1 and 2, we can prove Theorem 1 as follows:

Proof. Firstly, it is straightforward to verify that $\Phi^\lambda(v, L)$, the Social-Shapley value in graph-restricted games, is component efficient, additive and symmetric in the reason that both the Shapley value and the Solidarity value in cooperative games with graph structure satisfies these three properties. Note that for any

$\langle v, L\rangle \in \mathcal{G}_N^L$, $i \in N$ and $\alpha = \frac{1-\lambda}{\lambda} \geq 0$.

$$\begin{aligned}\Phi_i^\lambda(v, L) &= \sum_{i\in S} \frac{(n-s)!(s-1)!}{n!}[\lambda v^L(S) - \lambda v^L(S\backslash\{i\}) + (1-\lambda)A^{v^L}(S)]\\ &= \sum_{i\in S} \frac{(n-s)!(s-1)!}{n!(1+\alpha)}[v^L(S) - v^L(S\backslash\{i\}) + \alpha A^{v^L}(S)].\end{aligned}$$

If i is a α-counter contribution player, then $v^L(S) - v^L(S\backslash\{i\}) = -\alpha A^{v^L}(S)$, hence $\Phi_i^\lambda(v, L) = 0$.

Reciprocally, assuming that there is a value Ψ on $\mathcal{G}_N^L$ satisfying the above four properties, then Ψ is a linear mapping [5]. Obviously, $\Phi^\lambda(v, L)$ is also a linear mapping. Hence, for each game $\langle w_T, L\rangle$, we have $\Phi^\lambda(w_T, L) = \Psi(w_T, L)$. Thus, the equation $\Phi^\lambda(v, L) = \Psi(v, L)$ exists for each $\langle v, L\rangle \in \mathcal{G}_N^L$. Then we proved the uniqueness. □

4 A Class of the Social-Shapley Values in Graph-Restricted Games

In this section, we will axiomatically characterize the class of all possible values $\{\Phi^\lambda(v, L)|\lambda \in [0, 1]\}$. First of all, we introduce the graph-version of some properties.

- *Rationality*: Given $\langle v, L\rangle \in \mathcal{G}_N^L$, $S \subseteq N/L$ and $i \in S$, if $v^L(S) - v^L(S\backslash\{i\} \geq 0)$ and $A^{v^L}(S) \geq 0$, then $\Phi_i(v, L) \geq 0$.

The Social-Shapley values in graph-restricted games involves the Solidarity value and the Shapley value together. As a result, the axioms which are relative to these two values will play a critical role in the axiomatization.

- *Proportionality of Shapley Value*: If player i is A-null player in games $\langle v, L\rangle$, $\langle w, L\rangle \in \mathcal{G}_N^L$, then
$$\Phi_i(v, L)Sh_i(w, L) = \Phi_i(w, L)Sh_i(v, L).$$
- *Proportionality of Solidarity Value*: If player i is a null player in games $\langle v, L\rangle$, $\langle w, L\rangle \in \mathcal{G}_N^L$, then
$$\Phi_i(v, L)Sol_i(w, L) = \Phi_i(w, L)Sol_i(v, L).$$

For any $i \in N$, denote the linear space of all games $\langle v, L\rangle \in \mathcal{G}_N^L$ in which player i is a null player as $\mathcal{G}_N^0(i)$ and the linear space of all games $\langle v, L\rangle \in \mathcal{G}_N^L$ in which player i is A-null player as $\mathcal{G}_N^1(i)$, and put $\mathcal{G}_N(i) := \{\langle v, L\rangle \in \mathcal{G}_N^L | v^L(\{i\}) = 0\}$.

Lemma 3. *For each player $i \in N$, $\mathcal{G}_N(i)$ is the direct sum of $\mathcal{G}_N^0(i)$ and $\mathcal{G}_N^1(i)$, which means*

$$\mathcal{G}_N(i) = \mathcal{G}_N^0(i) \oplus \mathcal{G}_N^1(i).$$

Proof. See the proof in the Appendix.

On the basis of the above three properties and Lemma 3, we can depict the class of values $\{\Phi^\lambda(v, L) | \lambda \in [0, 1]\}$ as follow:

Theorem 2. *A value $\Phi(v, L)$ on $\mathcal{G}_N^L$ satisfies component efficiency, additivity, symmetry, rationality, proportionality of Shapley value and proportionality of Solidarity value if and only if there exists $\alpha \geq 0$ such that $\Phi(v, L) = \Phi^\lambda(v, L)$ with $\lambda = \frac{1}{1+\alpha}$.*

Proof. See the proof in the Appendix.

We put forward an example from a wireless network to illustrate the difference of the Shapley value, the Solidarity value and the Social-Shapley value in graph-restricted games.

Example 1. Assuming a partial wireless network is composed of three nodes (Fig. 2).

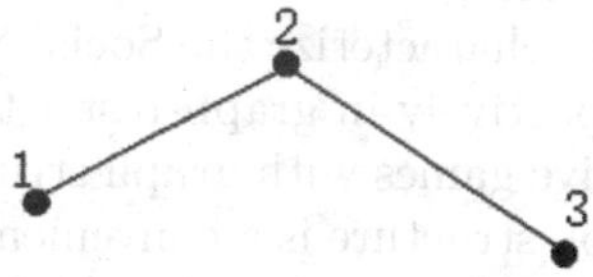

Fig. 2. A three nodes wireless network

Node 1 contributes nothing to any coalitions and Node 1 and Node 3 communicate through Node 2. Naturally, this situation can be condensed to a cooperative game with graph structure.

The graph-restricted game of these three nodes is $\langle N, v, L\rangle \in \mathcal{G}_N^L$, where $N = \{1, 2, 3\}$ and the graph $L = \{\{1, 2\}, \{2, 3\}\}$, $v(\emptyset) = 0$, $v(\{1\}) = 1$, $v(\{2\}) = 2$, $v(\{3\}) = 3$, $v(\{1, 2\}) = 3$, $v(\{1, 3\}) = 4$, $v(\{2, 3\}) = 8$, $v(\{1, 2, 3\}) = 9$.

Table 1 lists the calculation results of payoff distributions according to the Shapley value, the Solidarity value and the Social-Shapley value.

Note that, Node 1, as a dummy player, only gets his individual value by the Shapley value. In the second situation, Node 2 and Node 3, whose "bargaining power" are much stronger than that of Node 1, are quite generous to Node 1. The dummy player Node 1 gets $\frac{27}{12}$ according to the Solidarity value in graph-restricted game. The Social-Shapley value will adjust the sociality of a payoff through modifying the value of λ, which is depended on how friendly the players relation is. The smaller the coefficient λ is, the more support a dummy player can get. This support guards against the isolation of the dummy node by others in the wireless network to some extent. As a result, the Social-Shapley value $\Phi^\lambda(v, L)$ is much reasonable and acceptable in studying some wireless network problem and the social issues.

Table 1. The payoff distributions of three-node wireless network

		Node 1	Node 2	Node 3
$Sh(v,L)$		1	$\frac{7}{2}$	$\frac{9}{2}$
$Sol(v,L)$		$\frac{27}{12}$	$\frac{38}{12}$	$\frac{43}{12}$
$\Phi^{\lambda}(v,L)$	$\lambda=\frac{1}{5}$	2	$\frac{97}{30}$	$\frac{113}{30}$
	$\lambda=\frac{1}{2}$	$\frac{39}{24}$	$\frac{80}{24}$	$\frac{97}{24}$
	$\lambda=\frac{4}{5}$	$\frac{5}{4}$	$\frac{103}{30}$	$\frac{259}{60}$

5 Conclusions

According to the Shapley value, a player's payoff depends only on his own productivity measured by marginal contributions, which embodies utilitarianism in some sense. The Solidarity value reflects some social behavior of players in coalitions and indicates egalitarian. In order to balance the utilitarianism of the Shapley value and the egalitarian of the Solidarity value, we study the Social-Shapley value in graph-restricted games. In addition, we also offer corresponding graph-version of the axioms to characterize the Social-Shapley value and the class of Social-Shapley values respectively in graph-restricted games.

The research of cooperative games with graph structure has great theoretical significance because the graph structure is a convenient and intuitive demonstration of the partial cooperation relationships. Combining the properties of graph-restricted games, many solutions of cooperative games, such as the Banzhaf value [10], the EANS value [11], the CIS value [12] can be extended to the graph structure and be testified axiomatically.

Acknowledgement. This research has been supported by the National Natural Science Foundation of China (Grant No. 71571143), the Science and Technology Research and Development Program in Shaanxi Province of China (Grant Nos. 2017GY-095, 2017JM5147).

Appendix

Proof of Lemma 3

Given $i \in N$, let $\langle v, L\rangle = \mathcal{G}_N^0(i) \bigcap \mathcal{G}_N^1(i)$, then for each coalition $S \in C^L_{N\setminus\{i\}}$, we get $v^L(S\cup\{i\}) = v^(S)$ and $\sum_{k\in S\cup\{i\}}[v^L(S\cup\{i\}) - v^L(S\cup\{i\}\setminus\{k\})] = 0$. Thus, the following equation holds,

$$\sum_{k\in S}[v^L(S\cup\{i\}) - v^L(S\cup\{i\}\setminus\{k\})] = 0.$$

In this equation, let $S = \{j\}$, where $j \neq i$, we get $v^L(\{i,j\}) = v^L(\{i\})$. By $\langle v, L\rangle \in \mathcal{G}_N^0(i)$, we deduce $v^L(\{i\}) = 0$ and $v^L(\{i,j\}) = v^L(\{j\})$. Therefore, $v^L(\{j\}) = v^L(\{i,j\}) = v^L(\{i\}) = 0$.

Using mathematical induction on the cardinality of S, we can verify that $v^L(S) = v^L(S \cup \{i\}) = 0$ exists for each $S \in C^L_{N\setminus\{i\}}$, which means $\langle v, L\rangle$ is the null game, hence $\mathcal{G}^0_N(i) \bigcap \mathcal{G}^1_N(i) = \emptyset$.

Note that the dimension of the space $\mathcal{G}_N(i)$ equals $2^n - 2$. According to $\mathcal{G}^0_N(i) \subset \mathcal{G}_N(i)$, $\mathcal{G}^1_N(i) \subset \mathcal{G}_N(i)$, it remains to prove that the dimensions of the linear space $\mathcal{G}^0_N(i)$ and $\mathcal{G}^1_N(i)$ are at least $2^{n-1}-1$, respectively. Define two classes of games as

$$\mathcal{B}^0_N(i) = \{\langle u_T, L\rangle | i \notin T, T \neq \emptyset\}, \quad \mathcal{B}^1_N(i) = \{\langle w_T, L\rangle | i \notin T, T \neq \emptyset\},$$

where $\langle w_T, L\rangle$ is defined as in Sect. 3 and

$$u^L_T(S) = \begin{cases} 1, S = T \text{ or } S = T \cup \{i\}; \\ 0, \text{otherwise.} \end{cases}$$

Clearly, both $\mathcal{B}^0_N(i)$ and $\mathcal{B}^1_N(i)$ are composed of a set of linearly independent games.

Besides, since $\mathcal{B}^0_N(i) \subset \mathcal{G}^0_N(i)$, $\mathcal{B}^1_N(i) \subset \mathcal{G}^1_N(i)$ and the cardinalities of $\mathcal{B}^0_N(i)$ and $\mathcal{B}^1_N(i)$ are respectively $2^{n-1} - 1$, we can deduce the cardinalities of $\mathcal{G}^0_N(i)$ and $\mathcal{G}^1_N(i)$ are respectively $2^{n-1} - 1$. □

Proof of Theorem 2

It is easy to check that $\Phi^\lambda(v, L)$ satisfies component efficiency, additivity, symmetry, rationality, proportionality of Shapley value and proportionality of Solidarity value, the uniqueness will be proved as follow:

Step 1. We will prove for each $i \in N$, there exists $\lambda_i \in \mathbb{R}$, so that the following equality establish, $\langle v, L\rangle \in \mathcal{G}^L_N$,

$$\Phi_i(v, L) = \lambda_i Sh_i(v, L) + (1 - \lambda_i) Sol_i(v, L). \tag{3}$$

Choose any game $\langle w, L\rangle \in \mathcal{B}^1_N(i) \subset \mathcal{G}^1_N(i)$ such that $Sh_i(w, L) > 0$, let $\lambda_i = \frac{\Phi_i(w,L)}{Sh_i(w,L)}$, by proportionality of Shapley value, for all $\langle v, L\rangle \in \mathcal{G}^1_N(i)$, we have

$$\Phi_i(v, L) = \lambda_i Sh_i(v, L).$$

Similarly, choose any game $\langle u, L\rangle \in \mathcal{B}^0_N(i) \subset \mathcal{G}^0_N(i)$ such that $Sol_i(u, L) > 0$, let $\gamma_i = \frac{\Phi_i(u,L)}{Sol_i(u,L)}$, combining proportionality of Solidarity value, for each $\langle v, L\rangle \in \mathcal{G}^0_N(i)$, the following equality holds:

$$\Phi_i(v, L) = \gamma_i Sol_i(v, L).$$

According to the definitions of $\mathcal{G}^0_N(i)$ and $\mathcal{G}^1_N(i)$, given $\langle v, L\rangle \in \mathcal{G}^0_N(i)$, we have

$$\Phi_i(v, L) = \lambda_i Sh_i(v, L) + \gamma_i Sol_i(v, L). \tag{4}$$

Considering game $\langle w, L\rangle \in \mathcal{G}^L_N$, where

$$w^L(S) = \begin{cases} n, S = N; \\ 0, \text{otherwise,} \end{cases}$$

by component efficiency and symmetry, it follows $\Phi_i(w, L) = 1$, $i \in N$. Similarly, it follows that

$$Sh_i(w, L) = Sol_i(w, L) = 1, \ i \in N.$$

In the Eq. (4), let $\langle w, L\rangle = \langle v, L\rangle$, we get $\gamma_i = 1 - \lambda_i$. Consequently, the Eq. (3) holds for any $\langle v, L\rangle \in \mathcal{G}_N(i)$.

Now, defining a game $\langle u, L\rangle \in \mathcal{G}_N^L$ as follow:

$$u^L(S) = \begin{cases} 0, \, S = \emptyset; \\ 1, \, S \neq \emptyset. \end{cases}$$

Obviously $\langle u, L\rangle \notin \mathcal{G}_N(i)$.

Note that as linear space, the dimensions of $\mathcal{G}_N^L$ and $\mathcal{G}_N(i)$ are $2^n - 1$ and $2^n - 2$ respectively, hence, any $\langle v, L\rangle \in \mathcal{G}_N^L$ can be represented as follow

$$v = v^0 + cu,$$

where $\langle v^0, L\rangle \in \mathcal{G}_N(i)$ and c is a constant coefficient. According to component efficiency, additivity, symmetry, it follows that

$$\Phi_i(v, L) = \Phi_i(v^0, L) + \Phi_i(cu, L) = \lambda_i Sh_i(v^0, L) + (1 - \lambda_i) Sol_i(v^0, L) + \frac{c}{n}.$$

In addition, we have

$$\lambda_i Sh_i(v, L) + (1 - \lambda_i) Sol_i(v, L) = \lambda_i Sh_i(v^0, L) + (1 - \lambda_i) Sol_i(v^0, L) + \frac{c}{n}.$$

Consequently, Eq. (3) holds for any $\langle v, L\rangle \in \mathcal{G}_N^L$.

Step 2. We will prove that coefficient λ_i in Eq. (3) is independent of i. Assuming that $n \geq 2$, for $\langle v, L\rangle \in \mathcal{G}_N^L$, component efficiency implies:

$$\sum_{i \in N} [\lambda_i Sh_i(v, L) + (1 - \lambda_i) Sol_i(v, L)] = v^L(N).$$

By component efficiency of Solidarity value, this equation is equivalent to

$$\sum_{i \in N} \lambda_i [Sh_i(v, L) - Sol_i(v, L)] = 0.$$

By component efficiency of the Shapley and Solidarity value, we have

$$\begin{aligned} &\sum_{i=2}^{n} (\lambda_i - \lambda_1)[Sh_i(v, L) - Sol_i(v, L)] \\ &= \sum_{i \in N} \lambda_i [Sh_i(v, L) - Sol_i(v, L)] - \sum_{i \in N} \lambda_1 [Sh_i(v, L) - Sol_i(v, L)] \\ &= 0. \end{aligned}$$

This means for $\langle v, L\rangle \in \mathcal{G}_N^L$,

$$\sum_{i=2}^{n} \gamma_i [Sh_i(v, L) - Sol_i(v, L)] = 0, \tag{5}$$

where $\gamma_i = \lambda_i - \lambda_1$, $i = 2, 3, \ldots, n$.

Defining games $\langle v_k, L\rangle$, $k = 2, 3, \ldots, n$, as follow

$$v_k^L(S) = \begin{cases} 1, & S = N\backslash\{k\}; \\ 0, & \text{otherwise.} \end{cases}$$

According to Eq. (5),

$$\sum_{i=2}^{n} \gamma_i [Sh_i(v_k, L) - Sol_i(v_k, L)] = 0, \ k = 2, \ldots, n$$

and

$$Sh_i(v_k, L) = \begin{cases} -\frac{1}{n}, & i = k; \\ \frac{1}{n(n-1)}, & i \neq k, \end{cases}, \quad Sol_i(v_k, L) = \begin{cases} -\frac{1}{n^2}, & i = k; \\ \frac{1}{n^2(n-1)}, & i \neq k, \end{cases}$$

we get the linear system

$$(n-1)\gamma_k - \sum_{i \neq k} \gamma_i - 0, \ k = 2, 3, \ldots, n.$$

The only solution of the above linear system is $\gamma_2 = \gamma_3 = \cdots, \gamma_n = 0$.

Then for every $i \in N$, it implies $\lambda_i - \lambda_1 = 0$, $i = 2, 3, \ldots, n$. Now we have proved that coefficient λ_i is independent of i and $\lambda_i = \lambda_1 = \lambda$, $i = 2, 3, \ldots, n$.

Step 3. The coefficient λ will be proved belonging to interval $[0, 1]$ in this step.

Assuming that $n \geq 2$, for any player $i \in N$, considering the following unanimity games:

$$u_{N\backslash\{i\}}^L(S) = \begin{cases} 1, & S = N \text{ or } S = N\backslash\{i\}; \\ 0, & \text{otherwise.} \end{cases}$$

By applying the property of rationality to $\langle u_{N\backslash\{i\}}, L\rangle$, we obtain $\Phi_i(u_{N\backslash\{i\}}, L) \geq 0$. In addition, combining the following equations:

$$Sh_i(u_{N\backslash\{i\}}, L) = 0, \quad Sol_i(u_{N\backslash\{i\}}, L) = \frac{n-1}{n^2}, \ \lambda_i = \lambda,$$

we have

$$0 \leq \Phi_i(u_{N\backslash\{i\}}, L) = \lambda Sh_i(u_{N\backslash\{i\}}, L) + (1-\lambda) Sol_i(u_{N\backslash\{i\}}, L) = \frac{(1-\lambda)(n-1)}{n^2}.$$

Evidently, we deduce that $\lambda \leq 1$.

Considering the game $\langle \bar{w}, L\rangle$, which is defined as follow,

$$\bar{w}^L(S) = \begin{cases} -1, & S = N; \\ -n, & S = N\backslash\{i\}; \\ 0, & \text{otherwise.} \end{cases}$$

Applying the property of rationality to $\langle \bar{w}, L\rangle$, we get $\Phi_i(\bar{w}, L) \geq 0$. By the conditions:

$$Sh_i(\bar{w}, L) = \frac{n-1}{n}, \quad Sol_i(\bar{w}, L) = 0,$$

we have

$$0 \leq \Phi_i(\bar{w}, L) = \lambda Sh_i(\bar{w}, L) + (1-\lambda) Sol_i(\bar{w}, L) = \frac{\lambda(n-1)}{n},$$

which implies $\lambda \geq 0$. Then for $n \geq 2$, we have proven the conclusion. When $n = 1$, the theorem is obviously established. □

References

1. Myerson, R.B.: Graphs and cooperation in games. Math. Oper. Res. **2**(3), 225–229 (1977)
2. Li, X.: Algorithmic geometric and graphs issues in wireless networks. Wireless Commun. Mobile Comput. **3**(2), 119–140 (2010)
3. Chen, J., Lian, S.G., Fu, C., Du, R.Y.: A hybrid game model based on reputation for spectrum allocation in wireless networks. Comput. Commun. **33**(14), 1623–1631 (2010)
4. Shapley, L.S.: A value for n-person games. In: Kuhn, H.W., Tucker, A.W. (eds.) Contributions to the Theory of Games II. Annals of Mathematics Studies, pp. 307–317. Princeton, Princeton University Press (1953)
5. Nowak, A.S., Radzik, T.: A solidarity value for n-person transferable utility games. Int. J. Game Theor. **23**, 43–48 (1994)
6. Joosten, R.: Dynamics, Equilibria and Values, PhD Dissertation, Maastricht University (1996)
7. Casajus, A., Huettner, F.: Null players, solidarity, and the egalitarian Shapley values. J. Math. Econ. **49**, 58–61 (2013)
8. Nowak, A.S., Radzik, T.: On convex combinations of two values. Appl. Math. **24**(1), 47–56 (1996)
9. Ma, X., Sun, H.: Assignment for river's water resources based on solidarity value. Math. Pract. Theor. **43**, 131–137 (2013)
10. Dubey, P., Shapley, L.S.: Mathematical properties of the Banzhaf power index. Math. Oper. Res. **4**(2), 99–131 (1979)
11. Moulin, H.: The separability axiom and equal-sharing methods. J. Econ. Theor. **36**, 120–148 (1985)
12. Driessen, T.S.H., Funaki, Y.: Coincidence of and collinearity between game theoretic solutions. OR Spektrum **13**, 15–30 (1991)

The Extension of Combinatorial Solutions for Cooperative Games

Jiang-Xia Nan, Li-Xiao Wei, and Mao-Jun Zhang(✉)

School of Mathematics and Computing Science, Guangxi Colleges and Universities Key Laboratory of Data Analysis and Computation, Guilin University of Electronic Technology, Guilin 541004, Guangxi, China
zhang1977108@sina.com

Abstract. In this paper, a new model to solve the social selfish coefficient $\alpha \in [0, 1]$ for the α-egalitarian Shapley values and a new convex combinations of single-value in terms of a coalition forming weight coefficient $\beta \in [0, 1]$, which is called the SCE value for cooperative games are presented. The efficiency, linearity, symmetry and α-dummy player property of the SCE value are proved. By proposing a procedural interpretation, we define the β as the coalition forming weight (or possibility) coefficient and find a new way of assigning the grand coalition profit among all players is coincided with the SCE value which verify the SCE value's validity, applicability and superiority.

Keywords: Cooperative games · Egalitarian Shapley value · α-CIS value · SCE value

1 Introduction

Recently, with the rapid development of economy and society, the key problem of cooperative games with transferable utility is the distribution of the coalition's payoff among its players. A famous solution allots the grand coalition's profit based on the players' expected marginal contribution is the Shapley value [1] which is the embodiment of marginalism. A solution which equally divides the grand coalition's profit among all the players is the equal division (ED) value which is the representative of egalitarianism. The center of imputation set (CIS) value [2] is a solution which distributes the surplus of the overall profits to every player. It is the representative of utilitarianism. However, the above single value solutions always concern either individual or equity.

The convex combinations of the single-value solution which assume every player takes a certain percentage $\alpha \in [0,1]$ of a single-value solution and $1 - \alpha$ of another single-value solution give rise to new solution, which takes both utilitarianism and egalitarianism into account for cooperative game. Joosten [3] first introduced the α-egalitarian Shapley value by combining the Shapley value with ED value as a convex combinations with a fixed $\alpha \in [0, 1]$, and proved its efficiency, symmetry and linearity. Casajus [4] suggested the null player in a productive environment property to characterize the class of egalitarian Shapley values together with additivity, efficiency, and desirability. Brink et al. [5] proved that a solution satisfies efficiency, linearity

D. Li (Ed.): EAGT 2019, CCIS 1082, pp. 57–68, 2019.
https://doi.org/10.1007/978-981-15-0657-4_4

(or weak covariance), anonymity, and weak monotonicity if and only if it is an egalitarian Shapley value, and give a non-cooperative implementation to the class of values. Wang et al. [6] defined the α as social selfish coefficient and presented a procedural interpretation for the α-egalitarian Shapley value and characterize the α-egalitarian Shapley value by associated consistency, continuity and the α-dummy player property.

The CIS value is firstly defined by Driessen and Funaki [2] who also studied several conditions for the coincidence of the prenucleolus and the egalitarian non-separable contribution (ENSC) value and found that the Shapley value can be written as a convex combination of the ENSC value and the CIS value. Chun and Park [7] characterized the CIS value by efficiency, symmetry, strategic equivalence and population fair ranking. Dragan et al. [8] studied the collinear of Shapley value and some egalitarian division value including the CIS value on the zero-normalized game with the concept of average worth. Brink et al. [9] defined α-CIS value by convex combinations of CIS value and ED value, studied properties of α-CIS value and the relationship between α-CIS value and ENSC value. The convex combinations of the ED value and the CIS value reconcile two major economic allocation thoughts: egalitarianism and marginalism. Xu et al. [10] presented the α-individual rationality and α-dummy player property of the α-CIS value and discussed the noncooperative interpretation of the α-CIS value. Hou et al. [11] proposed the optimal compromise value and studied the relationship with the ENSC value and CIS value. Hu and Li [12] presented a non-singleton covariance property for the α-CIS value.

Except the α-egalitarian Shapley value and the α-CIS value, some other convex combinations are also provided. Brink and Funaki [13] studied the class of solutions consisting of all convex combinations of the CIS value, the ENSC value and the ED value. They provided several characterizations of this class of solutions, and some important properties of this solutions were given. Ju et al. [14] obtained the consensus value by a recursive two-side negotiation process and the comparison with Shapley value. They also defined α-consensus value by convex combinations of Shapley value and CIS value.

However, in existing research the coefficient α of convex combinations solution is given subjectively for cooperative game, which may be lead to some unreasonable allocation. In order to make the convex combination solutions more reasonable, a new model is presented to calculate the coefficient $\alpha \in [0, 1]$ of α-egalitarian Shapley value for cooperative games. This model can also be used to other convex combination solutions of cooperative games. Assuming that all players' social selfish coefficients are different and let α_i be the social selfish coefficient of player i, which is changeable and α_i-egalitarian Shapley value be the payoff of player i. All players reach a cooperation agreement to narrow the gap of α_i, which means they would like to cooperate with the others who are not much more selfish than themselves. Thus, we set the minimal variance of α_i as an objective function and solve the α_i under a lot of constraints including the rationality and efficiency. The solution shows that all players have the same social coefficient in cooperative games and give a reasonable mathematics interpretation of why all the players choose the same social selfish coefficient in cooperative games.

The convex combinations of single-value solutions provide new research direction for cooperative games, and its procedural interpretation makes the new solution reasonable for the reality. However, as mentioned above, Shapley value is the representative of marginalism, CIS value or ENSC value is the representative of utilitarianism and ED value is the representative of egalitarianism. The existing solutions based on convex combinations of single-value solution can only consider two major economic allocation thoughts. For example, α-egalitarian Shapley value which is the convex combinations of the equal ED value and the CIS value reconcile two major economic allocation thoughts: egalitarianism and marginalism. We also try to find a new solution based on the combinations of different single-value solutions. Thus, in this paper, we aim to propose a new convex combinations of single-value solutions called the SCE value, which is the linear combinations of Shapley value, CIS value and ED value and considers three major economic allocation thoughts simultaneously: marginalism, utilitarianism and egalitarianism.

The structure of this paper is organized as follows. In Sect. 2, we reviewer some single-value solutions and convex combinations of single-value solutions. In Sect. 3, a new model is presented to calculate selfish coefficients, which is to find selfish coefficient α_i in cooperative games under the constraints of rationality and efficiency. Section 4 presents the new convex combinations solution SCE value and discusses its property. In Sect. 5, the procedural interpretation of forming and diving the grand coalition is discussed to define the new coefficient β as a coalition forming weight (possibility) level and show the validity, applicability and superiority of the SCE value in cooperative games. Finally, the paper is concluded by a briefly summary in Sect. 6.

2 Preliminaries

Let $N = \{1, 2, \cdots, n\}$ be a non-empty, finite and fixed set. For any player $i \in N$ and any subset coalition $S \subseteq N$ of a transferable utility cooperative game paired by (N, v), its characteristic function $v : 2^N \rightarrow R$ is a real-valued function where the set of all the subsets of N is denoted by 2^N and $v(S)$ represents the worth of coalition S such that $v(\varnothing) = 0$. The class of all cooperative games with n-persons is denoted by $\mathcal{G}^N$ and $(N, v) \in \mathcal{G}^N$. A payoff vector of the game (N, v) is an $|N|$-dimensional real vector $x = (x_1, x_2, \cdots, x_n) \in R^N$, $n = |N|$, which represents a distribution of the payoffs that can be earned by cooperation over the individual players. Among all the single-value solutions for cooperative games, the Shapley value, CIS value and the ED value are essential and famous solutions.

Definition 1. [1] For any cooperative game $(N, v) \in \mathcal{G}^N$, and $\forall S \subseteq N$, the Shapley value is defined as follows:

$$Sh_i(N, v) = \sum_{\substack{S \subseteq N \\ i \in S}} \frac{(|N| - |S|)!(|S| - 1)!(v(S) - v(S \backslash \{i\}))}{|N|!}, \text{for all } i \in N \quad (1)$$

Where $|S|$ represents the number of the players in the coalition S and $v(S) - v(S \backslash \{i\})$ is the so-called marginal contribution of player i in coalition S.

Definition 2. [2] For any cooperative game $(N, v) \in \mathcal{G}^N$, the CIS value is defined by

$$CIS_i(N, v) = v(\{i\}) + n^{-1}\left[v(N) - \sum_{j \in N} v(\{j\})\right], \text{for all } i \in N. \tag{2}$$

Definition 3. For any cooperative game $(N, v) \in \mathcal{G}^N$, the ED value is given by

$$ED_i(N, v) = \frac{v(N)}{n}, \text{for all } i \in N. \tag{3}$$

Combining the Shapley value and ED, Joosten [3] proposes a new solution concept α-egalitarian Shapley value for cooperative games.

Definition 4. [3] For any cooperative game $(N, v) \in \mathcal{G}^N$, the α-egalitarian Shapley value is given by

$$\phi_i^\alpha(N, v) = \alpha Sh_i(N, v) + (1 - \alpha)ED_i(N, v), \text{for all } i \in N. \tag{4}$$

Another convex combinations of CIS value and ED value is presented by Brink et al. [9].

Definition 5 [9] For any cooperative game $(N, v) \in \mathcal{G}^N$, the α-CIS value is given by

$$\begin{aligned}\varphi_i^\alpha(N, v) &= \alpha CIS_i(N, v) + (1 - \alpha)ED_i(N, v) \\ &= \alpha v(\{i\}) + n^{-1}\left[v(N) - \sum_{j \in N} \alpha v(\{j\})\right], \text{ for all } i \in N\end{aligned} \tag{5}$$

The α-egalitarian Shapley value and α-CIS value of cooperative games both satisfy some well-known properties, which have been proved already. Let the solution of cooperative game (N, v) is denoted as $\eta_i(N, v)$, for any player $i \in N$, it follows that:

Efficiency: For any cooperative game $(N, v) \in \mathcal{G}^N$, $\sum_{i=1}^{n} \eta_i(N, v) = v(N)$.

Linearity: For any cooperative game (N, v), $(N, w) \in \mathcal{G}^N$, and $\forall a, b \in R$, $\eta_i(N, av + bw) = a\eta_i(N, v) + b\eta_i(N, w)$, where $av + bw$ is given by

$$(av + bw)(S) = v(S) + w(S), \text{ for all } S \subseteq N.$$

Symmetry: For any cooperative game $(N, v) \in \mathcal{G}^N$ and players $i, j \in N$, if $\eta_i(N, v) = \eta_j(N, v)$, then they are symmetric.

α-Dummy player property: For any cooperative game $(N, v) \in \mathcal{G}^N$, if player $i \in N$ is a dummy player, then $\eta_i(N, v) = \alpha v(\{i\}) + (1 - \alpha)\frac{v(N)}{n}$, $\alpha \in [0, 1]$.

3 The Model for the α-Egalitarian Shapley Values of Cooperation Games

In real world, every player may have its own selfish coefficient which can be alterable to meet with the cooperative requirement such as efficiency and rationality. In other words, the same α is too strict for cooperative games which leads to some unreasonable cooperation. When players form a grand coalition with various α_i, the key issue is how to choose their α_i to obtain acceptable payoffs under given coalitions characteristic functions.

On the basis of the definition of α as a social selfish coefficient, suppose that every player's social selfish coefficient α_i is different and all players can change their social selfish coefficient to get feasible allocations under the constraint of efficiency and rationality.

Definition 6. The α_i-egalitarian Shapley value is given by

$$\phi_i^{\alpha_i}(N,v) = \alpha_i Sh_i(N,v) + (1-\alpha_i)ED_i(N,v). \tag{6}$$

Actually, none of the Shapley value, the ED value and the α-egalitarian Shapley value is always satisfied with the rationality which requires all players allocation values are not less than their individual values and can be expressed as follows:

Rationality: For any cooperative game $(N,v) \in \mathcal{G}^N$ and $\forall\, i \in N$, $\phi_i^{\alpha}(N,v) \geq v(\{i\})$.

Given the assumption of different social selfish coefficient α_i, define a new allocation plan that let the variance of players, α_i be the discrepancy or fluctuation of their social selfish coefficient and the aim is to make the variance be minimum which means that a player would like to cooperate with those who are not selfish than itself.

Contracting the minimal variance model whose objective function is

$$\min Var(\alpha) = \frac{1}{n}\sum_{i\in N}\sum_{j\neq i, j\in N}(\alpha_i - \alpha_j)^2$$

or

$$\min \sum_{i=1}^{n}(\alpha_i - \bar{\alpha})^2 \ (\bar{\alpha} = \frac{\sum_{i\in N}\alpha_i}{n})$$

for the α_i-egalitarian Shapley values of cooperative games, which takes the rationality into account to obtain every α_i as follows:

$$\min z = Var(\alpha)$$
$$s.t. \begin{cases} \phi_i(N,v) = \alpha_i Sh_i(N,v) + (1-\alpha_i)ED_i \\ \phi_i(N,v) \geq v(\{i\}) \\ \sum_{i\in N}\phi_i(N,v) = v(N) \\ \alpha_i \in [0,1] \end{cases} \tag{7}$$

Let $\alpha^* = (\alpha_1^*, \alpha_2^*, \cdots, \alpha_n^*)$ be the solution vector of the above model (7) and $x^* = (x_1^*, x_2^*, \cdots, x_n^*)$ $(x_i^* = \phi_i^*(N, v))$ be the solution of the cooperative game. As all α_i^* have been obtained, the players' payoffs vector $\boldsymbol{x}^*$ in the grand coalition can be calculated accurately.

It is obviously that the optimal solution of model (7) makes the objective function $z = 0$, which means that all the α_i^* are equal and all the players have the same social selfish coefficients. The solution of this model can be defined as minimal variance α^* for egalitarian Shapley value short by MVES and the payoff of player i is

$$x_i^* = \phi_i^{\alpha^*}(N, w) = \alpha^* Sh_i(N, w) + (1 - \alpha^*) ED_i(N, w),$$

which is generated by the model (7).

Specially, it can be easily noticed that the optimal solution of the model (7) always shows that the social selfish level α_i among all the players turns out to be the same. That is to say, in cooperative games, the minimal variance min $z = Var(\alpha) \equiv 0$ because of the fact that $\sum_{i \in N} \phi_i^{\alpha}(N, v) \equiv v(N)$ when all players have the same social selfish coefficient α^*, it makes the efficiency is a useless constraint condition and no matter in which order the grand coalition is formed, the results of an agreement to narrow their gap of social selfish levels is that they should have the same α^*. The solution of the model (7) gives us an interpretation of why the social selfish coefficient α of all players should be the same.

Obviously, $\boldsymbol{x}^*$ satisfies the rationality and it satisfies the properties mentioned above.

Theorem 1. For any cooperative game $(N, v) \in \mathcal{G}^N$ and $\forall\, i \in N$, the MVES x_i^* satisfies rationality, efficiency, linearity, symmetry and α-dummy player property.

As the MVES is also one of the α-egalitarian Shapley value, it satisfies efficiency, linearity, symmetry an α-dummy player property. Moreover, the rationality condition has been included in the model, so the MVES also satisfies the rationality property.

To show the solution of model (7) specifically, the following example is given.

Example 1. Let v be the cooperative game defined by $v(1) = 30$; $v(2) = 25$; $v(3) = 40$; $v(12) = 70$; $v(13) = 90$; $v(23) = 80$; $v(123) = 150$; otherwise $v(S) = 0$.

Using Eqs. (1) and (3), the Shapley values and the ED values are:

$$Sh_1 = \frac{295}{6};\ Sh_2 = \frac{125}{3};\ Sh_3 = \frac{355}{6};\ ED_1 = ED_2 = ED_3 = 50.$$

Then, the α_i-egalitarian Shapley values are calculated by Eq. (6):

$$\phi_1 = \frac{295}{6}\alpha_1 + 50(1 - \alpha_1);\ \phi_2 = \frac{125}{3}\alpha_2 + 50(1 - \alpha_2);\ \phi_3 = \frac{355}{6}\alpha_3 + 50(1 - \alpha_3).$$

By model (7), the minimal variance model is constructed as follows:

$$\min z = (\alpha_1 - \bar{\alpha})^2 + (\alpha_2 - \bar{\alpha})^2 + (\alpha_3 - \bar{\alpha})^2$$

$$s.t.\begin{cases} \phi_1 = \frac{295}{6}\alpha_1 + 50(1-\alpha_1), \\ \phi_2 = \frac{125}{3}\alpha_2 + 50(1-\alpha_2), \\ \phi_3 = \frac{355}{6}\alpha_3 + 50(1-\alpha_3), \\ \phi_1 \geq 30, \\ \phi_2 \geq 25, \\ \phi_3 \geq 40, \\ \sum_i^3 \phi_i = 150, \\ \alpha_i \in [0,1]. \end{cases}$$

By calculation, we have $\alpha_1 = \alpha_2 = \alpha_3 = \alpha^* \approx 0.52$ and the objective function $z = 0$. The solving results of the players payoffs are $x_1 \approx 49.57$, $x_2 \approx 45.67$, $x_3 \approx 54.76$ and the payoff vector is $(49.57, 45.67, 54.76)$.

It can be seen from Example 1 that social selfish coefficient α and the corresponding α-egalitarian Shapley value can be obtained by the model (7), and the solution satisfies the individual rationality. Thus, the solution of this model shows a feasibility and applicability solution for all kinds of cooperative games.

4 A New Convex Combinations Solution of the Cooperative Games

The convex combinations of single-value solution create a new field for cooperative games. According to the α-egalitarian Shapley value and α-CIS value, we consider a new convex combinations of them, it follows:

$$\theta_i^{\alpha,\beta}(N,v) = \beta\phi_i^{\alpha}(N,v) + (1-\beta)\varphi_i^{\alpha}(N,v), \text{ for } \alpha,\ \beta \in [0,1].$$

As Wang et al. [6] and Hou [15] put it that the α represents the social selfish coefficient. Then the α is the same parameter in both the α-egalitarian Shapley value and α-CIS value. Thus according to Eqs. (4) and (5) we have

$$\begin{aligned} \theta_i^{\alpha,\beta}(N,v) &= \beta\phi_i^{\alpha}(N,v) + (1-\beta)\varphi_i^{\alpha}(N,v) \\ &= \beta[\alpha Sh_i(N,v) + (1-\alpha)ED_i(N,v)] \\ &\quad + (1-\beta)[\alpha CIS_i(N,v) + (1-\alpha)ED_i(N,v)] \\ &= \alpha\beta Sh_i(N,v) + \alpha(1-\beta)CIS_i(N,v) + (1-\alpha)ED_i(N,v) \end{aligned}$$

where the $\theta_i^{\alpha,\beta}(N,v)$ is the linear combinations of Shapley value, CIS value and ED value. Thus, it can consider three major economic allocation thoughts simultaneously: marginalism, utilitarianism and egalitarianism.

Definition 7. The SCE value of cooperative games is defined as follows:

$$\begin{aligned}\theta_i^{\alpha,\beta}(N,v) &= \beta\phi_i^{\alpha}(N,v) + (1-\beta)\varphi_i^{\alpha}(N,v)\\ &= \alpha\beta Sh_i(N,v) + \alpha(1-\beta)CIS_i(N,v)\\ &\quad + (1-\alpha)ED_i(N,v)\end{aligned} \tag{8}$$

where the $\alpha \in [0,1]$ is the social selfish coefficient and $\beta \in [0,1]$ is coalition forming weight coefficient.

Obviously, the SCE value contains three single-valued solutions and two parameters, thus, we can get its special form from the special values of two parameters α and β.

Property 1. For any cooperative game $(N,v) \in \mathcal{G}^N$ and $\forall\alpha,\beta \in [0,1]$, we have

(1) The SCE value is the Shapley value, when $\alpha = \beta = 1$, it follows that $Sh_i(N,v) = \theta_i^{1,1}(N,v)$.
(2) The SCE value is the CIS value, when $\alpha = 1, \beta = 0$, it follows that $CIS_i(N,v) = \theta_i^{1,0}(N,v)$.
(3) The SCE value is the ED value, when $\alpha = 0$, it follows that $ED_i(N,v) = \theta_i^{0,\beta}(N,v)$, $\beta \in [0,1]$.
(4) the SCE value $\theta_i^{\alpha,\beta}(N,v)$ can also be written as

$$\begin{aligned}\theta_i^{\alpha,\beta}(N,v) &= \alpha\beta Sh_i(N,v) + \alpha(1-\beta)CIS_i(N,v) + (1-\alpha)ED_i(N,v)\\ &= \alpha\beta\theta_i^{1,1}(N,v) + \alpha(1-\beta)\theta_i^{1,0}(N,v) + (1-\alpha)\theta_i^{0,\beta}(N,v)\end{aligned}$$

for $\alpha,\beta \in [0,1]$.

Based on Property 1, it is not difficult that we have the following Property 2.

Property 2. For any cooperative game $(N,v) \in \mathcal{G}^N$ and $\forall\alpha$, $\beta \in [0,1]$, it holds that

(1) $\theta_i^{\alpha,\beta}(N,v) + \theta_i^{\alpha,1-\beta}(N,v) = \theta_i^{\alpha,1}(N,v) + \theta_i^{\alpha,0}(N,v)$
(2) $\theta_i^{\alpha,\beta}(N,v) + \theta_i^{\delta,\beta}(N,v) = \theta_i^{\alpha+\delta,\beta}(N,v) + \theta_i^{0,\beta}(N,v)$ for all $\delta \in [0,1]$ such that $\alpha + \delta \in [0,1]$.

In addition, as the SCE value is the convex combinations of α-egalitarian Shapley value and α-CIS value which share the same properties of efficiency, linearity, symmetry and α-dummy player property. According to the operation rules of addition the SCE value also satisfies these properties.

Property 3. For any cooperative game $(N,v) \in \mathcal{G}^N$, $i \in N$, and $\forall\alpha$, $\beta \in [0,1]$, the SCE value satisfies efficiency, linearity, symmetry and α-dummy player property.

Note that the coefficient α of α-egalitarian Shapley value and α-CIS value has been given a reality meaning which is defined by the procedural interpretation by Xu et al. [10]. Similarly, we also try to find out the meaning of β to illustrate the possibility and the necessity of the SCE value in some game theory situations.

5 A Procedural Interpretation for the SCE Value

For any cooperative game $(N, v) \in \mathcal{G}^N$, let $\prod(N)$ be the set of all permutations on N. For any player $i \in N$ and any permutation $\pi \in \prod(N)$, denote $S_i^{\pi(k)} := \{\pi(k) \in N \,|k \leq \pi^{-1}(i)\}$ as the new coalition formed by arriving player in the order π and $P_i^{\pi(k)} := \{\pi(k) \in N | k \geq \pi^{-1}(i)\}$ as the original coalition of leaving player and his successors players in the same order π relatively, where $\pi^{-1}(i)$ denotes the order of player i in order π. In the following, we present- a procedure according to Wang et al. [6] and Hou et al. [15] in terms of the social selfish coefficient $\alpha \in [0, 1]$ and the coalition forming weight coefficient $\beta \in [0, 1]$, to generate a SCE value. The coalition forming weight coefficient β indicates the distribution weight of a coalition forming and $1 - \beta$ means dividing weight coefficient.

Step 1. The players arrive or leave in a random same order π, and all orders in $\prod(N)$ of arriving or leaving have the same possibility.
Step 2. Every player $i \in N$, joins into the coalition to form a new coalition $S_i^{\pi(k)}$,and the original coalition to leave to be $P_i^{\pi(k)}$, the first player takes $\alpha v(\{i\})$ no-matter he arrives or leaves, where the α is the social selfish coefficient $\alpha \in [0, 1]$.
Step 3. The arriving player $i \in N$, takes the payoff $\alpha v\left(S_i^{\pi(k)}\right)$ and have the surplus $v\left(S_i^{\pi(k)}\right) - v\left(S_i^{\pi(k)} \backslash \{i\}\right) - \alpha v\left(S_i^{\pi(k)}\right)$ shared by his successors averagely. When the player $i \in N$ is about to leave, he can take $\alpha v(\{i\})$ and the surplus $v\left(P_i^{\pi(k)}\right) - v\left(P_i^{\pi(k-1)}\right) - \alpha v(\pi(k))$ is divided equally by the remained players.
Step 4. The last arriving player $\pi(n)$ can only obtain his marginal contribution $v(N) - v(N \backslash \pi(n))$ and the last leaving player get $v(\pi(n))$, then the whole amount of $v(N)$ is distributed among all arriving players or leaving players in the order π.
Step 5. The grand coalition contains the procedures of both the forming and separating, let $\beta \in [0, 1]$ be the coalition forming weight coefficient (or possibility), oppositely, let $1 - \beta$ be the coalition diving weight coefficient,ϕ_i^π and φ_i^π be the arriving value and leaving value of player $i \in N$, and $\gamma_i^\pi = \beta\phi_i^\pi + (1 - \beta)\varphi_i^\pi$ be final payoff of the player $i \in N$.
Step 6. The value of a player $i \in N$ is the expected payoff of $\gamma_i^\pi(N, v)$ of $v(N)$ for all $\pi \in \prod(N)$.

For any cooperative game $(N, v) \in \mathcal{G}^N$ and $\forall \pi \in \prod(N)$, the payoff $\gamma_i^\pi(N, v)$ of the player $i \in N$ is determined by the Steps 1–6 as follows:

$$\gamma_i^\pi = \begin{cases} \alpha v(\{i\}), & \pi^{-1}(i) = 1; \\ \beta\left[\alpha v\left(S_i^{\pi(k)}\right) + \lambda_i^\pi\right] + (1-\beta)\left[\alpha v(\{i\}) + \mu_i^\pi\right], & 1 < \pi^{-1}(i) < n; \\ \beta\left[v(N) - v(N \backslash \{i\}) + \lambda_i^\pi\right] + (1-\beta)\left[v(\{i\}) + \mu_i^\pi\right], & \pi^{-1}(i) = n; \end{cases}$$

Where

$$\lambda_i^\pi = \sum_{k=1}^{\pi^{-1}(i)-1} \frac{v\left(S_i^{\pi(k)}\right) - v\left(S_i^{\pi(k)} \backslash \pi(k)\right) - \alpha v\left(S_i^{\pi(k)}\right)}{n-k}$$

and

$$\mu_i^\pi = \sum_{k=1}^{\pi^{-1}(i)-1} \frac{v\left(P_i^{\pi(k)}\right) - v\left(P_i^{\pi(k)} \backslash \pi(k)\right) - \alpha v(\pi(k))}{n-k}.$$

The value $\gamma_i(N, v)$ defined in the above procedure is the expected payoff of $\gamma_i(N, v)$ in Step 6 as follows:

$$\gamma_i(N, v) = \frac{1}{n!} \sum_{\Pi(N)} \gamma_i^\pi(N, v),$$

for all $i \in N$.

Let π^* be the reverse order of π, the leaving procedure of player $i \in N$ is the reverse procedural interpretation of the one studied by Hou et al. [15]. For any $\pi \in \prod(N)$, it holds that $\pi^* \in \prod(N)$. Every order $\pi \in \prod(N)$ must have a reverse order π^* which means the expected payoff of φ_i^π coincide with the bargaining outcome presented by Hou et al. [15].

By the proof of Wang et al. [6] and Hou et al. [15], Theorem 2 can be obtained that:

Theorem 2. For any cooperative game $(N, v) \in \mathcal{G}^N$ and for any α-egalitarian Shapley value $\phi_i^\alpha(N, v)$, α-CIS value $\varphi_i^\alpha(N, v)$, The following equation holds:

$$\phi_i^\alpha(N, v) = \frac{1}{n!} \sum_{\Pi(N)} \phi_i^\pi(N, v)$$

$$\varphi_i^\alpha(N, v) = \frac{1}{n!} \sum_{\Pi(N)} \varphi_i^\pi(N, v)$$

The above expected payoff value actually is in accordance with the SCE value and the procedure is the interpretation for the SCE value.

Theorem 3. For any cooperative game $(N, v) \in \mathcal{G}^N$, the expected value $\gamma_i(N, v)$ is in accordance with the SCE value $\theta_i^{\alpha,\beta}(N, v)$.

Proof. From Theorem 2, we have

$$\begin{aligned}
\gamma_i(N,v) &= \frac{1}{n!}\sum_{\Pi(N)} \gamma_i^{\pi}(N,v) \\
&= \frac{1}{n!}\sum_{\Pi(N)} \left[\beta\phi_i^{\pi}(N,v) + (1-\beta)\varphi_i^{\pi}(N,v)\right] \\
&= \beta\frac{1}{n!}\sum_{\Pi(N)} \phi_i^{\pi}(N,v) + (1-\beta)\frac{1}{n!}\sum_{\Pi(N)} \varphi_i^{\pi}(N,v) \\
&= \beta\phi_i^{\alpha}(N,v) + (1-\beta)\varphi_i^{\alpha}(N,v) \\
&= \theta_i^{\alpha,\beta}(N,v)
\end{aligned}$$

On the basis of the procedure interpretation above, the α-egalitarian Shapley value shows the result of arriving the grand coalition in all order $\pi \in \prod(N)$ while the α-CIS value is the payoff of leaving the grand coalition in all order $\pi \in \prod(N)$ which interpret the two convex combinations in a new way and explain the significance of assigning the payoff of $v(N)$ by using the SCE value.

6 Concluding Remarks

In this paper, a model to solve the social selfish coefficient $\alpha \in [0, 1]$ for the α- egalitarian Shapley values and a new convex combination of single-value called SCE value are presented. Instead of the existing methods for giving selfish coefficient α, we present a minimal variance model to find the coefficient α for α-egalitarian Shapley value which assume different α_i and the accordance of narrowing the gap of α_i. By model (7), social selfish coefficient α and the corresponding α-egalitarian Shapley value can be obtained, and the solution satisfies the individual rationality. In addition, the SCE value which is the convex combinations of two convex combinations solutions α-egalitarian Shapley value and α-CIS value is discussed. The SCE value satisfies the same property efficiency, linearity, symmetry, and α-dummy player property. The procedure of the forming and separating of a grand coalition can explain how the α-egalitarian Shapley value and α-CIS value are obtained in which the definition of coalition forming coefficient $\beta \in [0, 1]$ makes the SCE value be considers three major economic allocation thoughts simultaneously: marginalism, utilitarianism and egalitarianism.

Acknowledgement. The authors would like to thank the associate editor and also appreciate the constructive suggestions from the anonymous referees. This research was supported by the key Program of National Natural Science Foundation of China (No. 71231003), the Natural Science Foundation of China (Nos. 71461005 and 71561008). The Innovation Project of Guet Graduate Education (No. 2019YCXS082).

References

1. Shapley, L.S.: A value for n-person games. Ann. Math. Stud. **28**(7), 307–318 (1953)
2. Driessen, T.S.H., Funaki, Y.: Coincidence of and collinearity between game theoretic solutions. Oper. Res. Spektrum **13**(1), 15–30 (1991)
3. Joosten, R.: Dynamics, equilibria and values. Ph.D. dissertation, Maastricht University (1996)
4. Casajus, A., Huettner, F.: Null players, solidarity, and the egalitarian Shapley values. J. Math. Econ. **49**(1), 58–61 (2013)
5. Brink, R., Funaki, Y., Ju, Y.: Reconciling marginalism with egalitarianism: consistency, monotonicity, and implementation of egalitarian Shapley values. Soc. Choice Welfare **40**(3), 693–714 (2013)
6. Wang, W.N., Sun, H., Xu, G.J., Hou, D.S.: Procedural interpretation and associated consistency for the egalitarian Shapley values. Oper. Res. Lett. **45**(2), 164–169 (2017)
7. Chun, Y., Park, B.: Population solidarity, population fair-ranking, and the egalitarian value. Int. J. Game Theory **41**(2), 255–270 (2012)
8. Dragan, I., Driessen, T., Funaki, Y.: Collinearity between the Shapley value and the egalitarian division rules for cooperative games. Oper. Res. Spektrum **18**(2), 97–105 (1996)
9. Brink, R., Chun, Y., Funaki, Y., Park, B.: Consistency, population solidarity, and egalitarian solutions for TU-Games. Theor. Decis. **81**(3), 427–447 (2016)
10. Xu, G.J., Dai, H., Shi, H.B.: Axiomatizations and a Noncooperative Interpretation of the α-CIS Value. Asia-Pac. J. Oper. Res. **32**(5), 1550031 (2015)
11. Hou, D.S., Sun, P.F., Xu, G.J., Driessen, T.: Compromise for the complaint: an optimization approach to the ENSC value and the CIS value. J. Oper. Res. Soc. **63**(3), 1–9 (2017)
12. Hu, X.F.: A new axiomatization of a class of equal surplus division values for TU games. RAIRO Oper. Res. **52**(3), 935–942 (2018)
13. Brink, R., Funaki, Y.: Axiomatizations of a class of equal surplus sharing solutions for TU-games. Theor. Decis. **67**(3), 303–340 (2009)
14. Ju, Y., Borm, P., Ruys, P.: The consensus value: a new solution concept for cooperative games. Soc. Choice Welfare **28**(4), 685–703 (2004)
15. Hou, D.S., Sun, H., Xu, G.J.: Compromise for the complaint: process and optimization approach to the alpha-CIS value. Working paper

An Allocation Value of Cooperative Game with Communication Structure and Intuitionistic Fuzzy Coalitions

Jie Yang(✉)

College of Management, Fujian Agriculture and Forestry University, Fuzhou 350002, Fujian, China
yangjie@fafu.edu.cn

Abstract. At present, the researches on the cooperative games mainly based on the hypothesis that arbitrary coalitions can be formed and the fuzzy coalitions are Aubin's form. However, it is always not true in reality. This paper defines a cooperative game with communication structure and intuitionistic fuzzy coalition, in which the partners have some hesitation degrees and different risk preferences when they take part in limited coalitions. There are lower and upper participation degrees of players in coalitions by introducing confidence levels to intuitionistic fuzzy coalitions. Then a formula of average tree solution (short called AT solution) for this cooperative game is proposed based on the defined preference weighted form by taking account of players' risk preferences, and the existence of the solution according to axioms system is proved. Finally, the effectiveness of this method is demonstrated by a practical example of profit allocation. This research extends the cooperative game with fuzzy coalitions, and integrates individual preferences information of players in cooperation.

Keywords: Graph cooperative games · Fuzzy coalition · Average tree solution · Intuitionistic fuzzy number · Risk preference

1 Introduction

In real life, cooperation is always limited due to the limitation of resources, status and culture etc. In another words, the coalition formation is not always existing among all players, which has some restrictions. For this situation, Myerson [1] defined a cooperative game with limited communication structure, in which vertices represent the players in a game and edges are the communication paths among them. The model was called a graph cooperative game. In this game, alliances exist if and only if the players can connect. He defined the Myerson value, and showed that the value lies in the core when the game is convex. The Myerson value is analogous to the Shapley value of a game restricted by a communication structure.

Since then some researchers also found that the cooperative games with limited communication structure can effectively reflect complicated cooperation situations, and they proposed corresponding solutions to distribution revenue [2–4]. An average tree solution (AT solution for short) was defined [5] for an acyclic graph cooperative game, and proved that the solution satisfies component efficiency and component fairness.

D. Li (Ed.): EAGT 2019, CCIS 1082, pp. 69–84, 2019.
https://doi.org/10.1007/978-981-15-0657-4_5

Component fairness means that if the link between two players is deleted, the change of allocation is same for both, which implies their marginal contributions in this component are equal. And an AT solution and a sub-cores of a graph cooperative game were defined, and studied whether the AT solution lies within the sub-core. Talman and Yamamoto [6] continued the analysis of the average tree solution by presenting a weaker condition than super-additivity, such that the average tree solution belongs to the core. Later, Herings et al. [7] generalized the AT solution for an acyclic graph cooperative game with communication structure. And a version of AT solution appropriate for a game with restricted coalition structure based on an acyclic graph [8]. The AT solution of graph cooperative game was concerned because its good properties [9], such as it must lie in the core if the game is super-additive but the Myerson value may not, and it is much simpler to find the marginal eigenvector of the allocation.

In addition, as we all known, players often are not able to evaluate situations exactly because imprecise or lack of available information in real game. Fuzzy cooperative games have attracted many researchers' attention because fuzzy models are needed to describe uncertain information in cooperation [10–13]. Aubin [10] introduced a fuzzy cooperative game with fuzzy coalitions to extend crisp coalitions, in which the degree of participation is a real number in the interval [0, 1]. In graph cooperative game, Jiménez-Losada et al. [11] defined fuzzy Myerson values by using the proportional model and the Choquet model. Xu et al. [12] proposed a Myerson value of cooperative games with communication structure and fuzzy coalitions. The common feature of these fuzzy coalitions is that the fuzziness is described by the fuzzy set, which the non-membership degree is just automatically equal to the complement of the membership degree to 1. In practice, however, players often do not express the non-membership degree of a given element as the complement of the membership degree. In other words, the players may have some hesitation degrees. The fuzzy set has no means to incorporate the hesitation degree. For example, a player just knows his/her degree in a coalition is 0.3 at least, and 0.5 must not in it, then his/her hesitation degree is 0.2. In order to make fuzzy cooperative game theory more applicable to real problems, the intuitionistic fuzzy set [14, 15] was introduced, which can express more abundant information by using membership degree, non-membership degree and hesitation degree.

Besides the above discussed, the risk attitudes of players should be considered in cooperation on account of the players are not identical in reality. Differing risk attitudes are well studied in the field of decision-making and non-cooperative games, such as [16–18], but this matter is rarely considered in cooperative games. In this paper, we identify with the players have different risk preferences, and study how to integrate individual preferences into cooperative game. In this paper, we will study the cooperative game with communication structure and intuitionistic fuzzy coalition based on preference weighted form.

This paper is organized as follows. Section 2, we define a graph cooperative game with intuitionistic fuzzy coalitions in view of the crisp graph cooperative game, and give the characteristic function by Choquet integral form. Section 3, a mean value of preference was defined, then a formula for the AT solution with Intuitionistic fuzzy coalitions was proposed, and the existence and rationality of the solution is proved by according to axioms system. The model and method proposed in this paper are illustrated by a profit allocation problem in Sect. 4. Conclusions appear in last Section.

2 Graph Cooperative Game with Intuitionistic Fuzzy Coalitions

2.1 Crisp Graph Cooperative Game and Its AT Solution

In a crisp cooperative game with transferable utility (a TU-game), the triad (N, v, L) constitutes a graph cooperative game, where $N = \{1, 2, \cdots, n\}$ is the set of players, $v : 2^N \to R$ is a mapping defined on subsets of N, and $L \subseteq \{\{i,j\} | i \neq j, i, j \in N\}$ is a set of links in the communication graph. The function v is the payoff (characteristic) function of a game, where $v(S)$ is the worth of each coalition $S \subseteq 2^N$, with the property $v(\varnothing) = 0$. The set of all games is denoted by $G(N)$.

If $L = \{\{i,j\} | i \neq j, i, j \in N\}$, then players in the game can ally with each other arbitrarily, and the corresponding graph is a complete graph (N, L). In this case, (N, v, L) is a cooperative game with complete communication structure or a complete graph cooperative game, and usually the well-known cooperative game refers to this one, which can be abbreviated as (N, v). In the game, any player in the game can form coalitions with others freely and unrestrictedly. If $L \neq \{\{i,j\} | i \neq j, i, j \in N\}$ and L is non-null, then (N, v, L) is a cooperative game with limited communication structure. In this game, players may have a cooperation if and only if they are interconnected, and coalitions forming is restrictedly. This paper discusses a cooperative game with cycle-free and restricted communication structure.

For an undirected graph (N, L), a coalition of players $K \subseteq N$ is a network of (N, L) if K is connected, i.e. between any two members of K, there is a path with in L. A network is called a component if no larger network contains it. We denote by $\hat{C}^L(N)$ the set of all components of (N, L). An n-tuple $B = (B_1, \cdots, B_n)$ of subsets of N is admissible if (1) For all $i \in N$, $i \in B_i$; (2) For some $j \in N$, $B_j = N$; (3) For all $i \in N$ and $K \in \hat{C}^L(B_i \backslash \{i\})$, there exists $j \in N$ such that $K = B_j$ and $\{i,j\} \in L$. In a crisp graph cooperative game (N, v, L), a AT solution is defined as follows [7]:

$$AT_i(N, v, L) = \frac{1}{|B^L|} \{ \sum_{B \in B^L} [v(B_i) - \sum_{K \in \hat{C}^L(B_i \backslash \{i\})} v(K)] \} \tag{1}$$

where $i = 1, 2, \cdots, n$, B^L is the collection of all admissible n-tuple of coalitions B, $|B^L|$ represents the number of components of B^L.

It has been proved that AT solution is characterized by efficiency, dummy, linearity, strong symmetry, and it satisfies properties of component efficiency, component fairness, and additivity [9].

For any component $K \in \hat{C}^L(N)$, $AT(N, v, L)$ satisfies component efficiency if it holds that

$$\sum_{i \in K} AT_i(N, v, L) = v(K)$$

For any component $K \in \hat{C}^L(N)$, $AT(N, v, L)$ satisfies component fairness if it holds that

$$\frac{1}{|K^h|}\sum_{i\in K^h}(AT_i(N,v,L) - AT_i((N,v,L)\backslash L\{h,l\})) = \frac{1}{|K^l|}\sum_{i\in K^l}(AT_i(N,v,L) - AT_i((N,v,L)\backslash L\{h,l\}))$$

where K^h and K^l express the subset of K of players connected to h and l in the graph that results after deleting the link $L\{h,l\}$. Component fairness says that the allocation change for player i in components K^h and K^l are the same if the edge $L\{h,l\}$ is deleted.

Let $(N, v_1, L) \in G(N)$ and $(N, v_2, L) \in G(N)$ be crisp graph cooperative games. If any $i \in N$,

$$AT_i(N, v_1 + v_2, L) = AT_i(N, v_1, L) + AT_i(N, v_2, L)$$

then the n-dimensional allocation vector $AT(N, v, L)$ satisfies additivity.

2.2 Fuzzy Coalition and Intuitionistic Fuzzy Coalition

In a cooperative game with fuzzy coalitions, the set N composed by all the fuzzy subsets is denoted as $F(N)$, the arbitrary element $\bar{S}$ represents the fuzzy coalition and could be demonstrated by fuzzy vectors as:

$$\bar{S} = (\bar{S}(1), \bar{S}(2), \cdots, \bar{S}(n)) : F(N) \to \mu_{\bar{S}}(i)$$

where $\mu_{\bar{S}}(i)$ is the degree of participation of the player i ($i = 1, 2, \cdots, n$) in the coalition $\bar{S}$, namely the ratio of the resources invested by player i to the total resources. Currently, the fuzzy coalition $\mu_{\bar{S}}(i)$ is described a real number in the interval [0, 1], and this kind of fuzzy coalitions $[0, 1]^n$ of cooperative games are still crisp numbers. In order to describe real uncertain situations by fuzzy numbers, and to express players' hesitation degree, the intuitionistic fuzzy set is considered as an effective tool for modeling the cooperative games.

Definition 1. [14] Let X be a nonempty set of the universe. If two mapping on the set X are $\mu_{\tilde{A}} : X \to [0, 1]$ and $\upsilon_{\tilde{A}} : X \to [0, 1]$, so that $0 \le \mu_{\tilde{A}}(x) \le 1$, $0 \le \upsilon_{\tilde{A}}(x) \le 1$ and $0 \le \mu_{\tilde{A}}(x) + \upsilon_{\tilde{A}}(x) \le 1$ $(x \in X)$, then $\mu_{\tilde{A}}$ and $\upsilon_{\tilde{A}}$ are called an intuitionistic fuzzy set X, denoted by $\tilde{A} = \{<x; \mu_{\tilde{A}}(x), \upsilon_{\tilde{A}}(x)> |x \in X\}$, where $\mu_{\tilde{A}}$ and $\upsilon_{\tilde{A}}$ are the membership degree and non-membership degree of $\tilde{A}$ respectively. The set of the intuitionistic fuzzy sets on the universal set X is denoted by $IF(X)$.

It is easy to see from Definition 1 that an intuitionistic fuzzy set is defined by a pair of membership and non-membership degrees (or functions), which are more or less independence of each other, and the sum of the membership degree and non-membership degree is not greater than 1. In a fuzzy coalition, the membership degree can express a player's level of participation into a coalition, while the non-membership degree is the level of not participation, and the rest is his/her hesitation degree.

Based on the definition of intuitionistic fuzzy set, we can generalize the cooperative game with fuzzy coalition set $F(N)$ into the cooperative game with intuitionistic fuzzy coalition set $IF(N)$. For any intuitionistic fuzzy coalition element $\tilde{S}$, $\mu_{\tilde{S}}(i) \in [0,1]$ is a participation degree of player $i(i = 1, 2, \cdots, n)$ in the coalition $\tilde{S}$, $\upsilon_{\tilde{S}}(i) \in [0,1]$ is non-participation degree, and $0 \leq \mu_{\tilde{S}}(x) + \upsilon_{\tilde{S}}(x) \leq 1$. Then the hesitation degree of player i is $\pi_{\tilde{S}}(i) = 1 - \mu_{\tilde{S}}(x) - \upsilon_{\tilde{S}}(x)$. We can express intuitionistic fuzzy coalition is $\tilde{S} = \{<1, \mu_{\tilde{S}}(1), \upsilon_{\tilde{S}}(1)>, <2, \mu_{\tilde{S}}(2), \upsilon_{\tilde{S}}(2)>, \cdots, <n, \mu_{\tilde{S}}(n), \upsilon_{\tilde{S}}(n)>\}$ by the representation way of intuitionistic fuzzy sets. Specially, if $\tilde{S} = \{<1;0,0>, <2;0,0>, \cdots, <n,0,0>\}$, then the alliance is empty; and $\tilde{S} = \{<1;1,0>, <2;1,0>, \cdots, <n;1,0>\}$ is a crisp grand coalition. Obviously, if $\mu_{\tilde{S}}(x) + \upsilon_{\tilde{S}}(x) = 1$, the intuitionistic fuzzy coalition becomes a fuzzy coalition.

Definition 2. For a finite set of players $N = \{1, 2, \cdots, n\}$, if the fuzzy payoffs $\tilde{v}$ of $(N, \tilde{v}, L)$ are mapping functions from intuitionistic fuzzy coalitions $\tilde{F}(N)$ to a fuzzy number set $\tilde{R}$, namely, $\tilde{v} : IF(N) \to \tilde{R}$ with $\tilde{v}(\varnothing) = 0$, then $(N, \tilde{v}, L)$ is a graph cooperative game with intuitionistic fuzzy coalitions. For conciseness, the entirety of graph cooperative games with intuitionistic fuzzy coalitions is denoted $G_{IF}(N)$. ∎

This paper discusses the most common graph cooperative game, which satisfies general properties of convexity and super additivity.

Definition 3. Let $(N, \tilde{v}, L)$ be a graph cooperative game with intuitionistic fuzzy coalitions. A coalition $\tilde{T} \in IF(N)$ is a carrier of $(N, \tilde{v}, L)$ if and only if it satisfies

$$\tilde{v}(\tilde{S} \cap \tilde{T}) = \tilde{v}(\tilde{S})$$

for any coalitions $\tilde{S} \in IF(N)$ with $\tilde{S} \cap \tilde{T} = \varnothing$.

Definition 4. Let $(N, \tilde{v}, L)$ be a graph cooperative game with intuitionistic fuzzy coalitions $(N, \tilde{v}, L)$, it is said to be convex if it satisfies

$$\tilde{v}(\tilde{S} \cup \tilde{T}) \geq \tilde{v}(\tilde{S}) + \tilde{v}(\tilde{T}) - \tilde{v}(\tilde{S} \cap \tilde{T})$$

for any coalitions $\tilde{S}, \tilde{T} \in IF(N)$ with $\tilde{S} \cap \tilde{T} = \varnothing$.

Definition 5. Let $(N, \tilde{v}, L)$ be a graph cooperative game with Intuitionistic fuzzy coalitions $(N, \tilde{v}, L)$, it is called super-additive if it satisfies

$$\tilde{v}(\tilde{S} \cup \tilde{T}) \geq \tilde{v}(\tilde{S}) + \tilde{v}(\tilde{T})$$

for any coalitions $\tilde{S}, \tilde{T} \in IF(N)$ with $\tilde{S} \cap \tilde{T} = \varnothing$.

In a $(N, \tilde{v}, L)$, the really participation degrees of players may be different by the attitude of hesitation degree. If the hesitation degree is supposed to take participate in the cooperation, then the maximum participation degrees is $\mu_{\tilde{S}}(i) + \pi_{\tilde{S}}(i)$, and the minimum participation degrees is $\mu_{\tilde{S}}(i)$.

Therefore, the participation degree of player $i(i \in N)$ could be defined as a closed interval number with a confidence level $\lambda(\lambda \in [0,1])$:

$$\eta_{\tilde{S}(\lambda)}(i) = [\mu_{\tilde{S}}(i), \mu_{\tilde{S}}(i) + \lambda\pi_{\tilde{S}}(i)] \tag{2}$$

where the lower level of participation is $\eta^-_{\tilde{S}}(i) = \mu_{\tilde{S}}(i)$, and the upper level of participation is $\eta^+_{\tilde{S}}(i) = \mu_{\tilde{S}}(i) + \pi_{\tilde{S}}(i) = 1 - \upsilon_{\tilde{S}}(i)$. It is clear that $0 \le \eta^-_{\tilde{S}}(i) \le \eta^+_{\tilde{S}}(i) \le 1$ $(\eta_{\tilde{S}}(i) \subseteq [0,1])$ for any player in the intuitionistic fuzzy coalition $\tilde{S} \in G_{IF}(N)$.

Definition 6. In a graph cooperative game with intuitionistic fuzzy coalitions $(N, \tilde{v}, L)$, for any coalition $\tilde{S} \in IF(N)$ with confidence level λ, there exists arbitrary $i \in \mathrm{Supp}(\tilde{S})$, $\mathrm{Supp}(\tilde{S}) = \{i \in N \,|\, \eta_{\tilde{S}(\lambda)}(i) > 0\}$, $D(\tilde{S}(\lambda)) = \{\eta_{\tilde{S}(\lambda)}(i) \,|\, \eta_{\tilde{S}(\lambda)}(i) > 0, i \in N\}$, and $d(\tilde{S})$ is the number of elements in $D(\tilde{S})$. If the elements in $D(\tilde{S})$ are arranged in an ascending order $0 < h^-_{1(\lambda)} \le h^-_{2(\lambda)} \cdots \le h^-_{d(\tilde{S}(\lambda))} \le 1$ and $0 < h^+_{1(\lambda)} \le h^+_{2(\lambda)} \cdots \le h^+_{d(\tilde{S}(\lambda))} \le 1$, then the characteristic function by Choquet integral form [19, 20] of $(N, \tilde{v}, L)$ can be expressed as:

$$\begin{aligned} \int \tilde{S}_\lambda dv &= \sum_{m=1}^{d(\tilde{S})} v([\tilde{S}]_{h_{m(\lambda)}})(h_{m(\lambda)} - h_{m-1(\lambda)}) \\ &= \sum_{m=1}^{d(\tilde{S})} [v([\tilde{S}]_{h^-_{m(\lambda)}})(h^-_{m(\lambda)} - h^-_{m-1(\lambda)}), v([\tilde{S}]_{h^+_{m(\lambda)}})(h^+_{m(\lambda)} - h^+_{m-1(\lambda)})] \end{aligned} \tag{3}$$

where $[\tilde{S}]_{h^-_{m(\lambda)}} = \{i \in N | \eta^-_{\tilde{S}(\lambda)}(i) = h^-_{m(\lambda)}\}$ and $[\tilde{S}]_{h^+_{m(\lambda)}} = \{i \in N | \eta^+_{\tilde{S}(\lambda)}(i) = h^+_{m(\lambda)}\}$ are the fuzzy coalitions composed by all the players whose participation degrees are $\eta^-_{\tilde{S}(\lambda)}(i) = h^-_{m(\lambda)}$ and $\eta^+_{\tilde{S}(\lambda)}(i) = h^+_{m(\lambda)}$ respectively, $v([\tilde{S}]_{h^-_{m(\lambda)}})$ and $v([\tilde{S}]_{h^+_{m(\lambda)}})$ are the payoffs of a graph cooperative game with crisp coalitions.

According to Eq. (3), the characteristic functions of $(N, \tilde{v}, L)$ are interval numbers because the fuzzy coalitions of the graph cooperative game are interval values by Eq. (2), whose maximum and minimum are respectively denoted as follows:

$$tv^-_\lambda(\tilde{S}) = \sum_{m=1}^{d(\tilde{S})} v([\tilde{S}]_{h^-_{m(\lambda)}})(h^-_{m(\lambda)} - h^-_{m-1(\lambda)}), tv^+_\lambda(\tilde{S}) = \sum_{m=1}^{d(\tilde{S})} v([\tilde{S}]_{h^+_{m(\lambda)}})(h^+_{m(\lambda)} - h^+_{m-1(\lambda)}) \tag{4}$$

where $h(0) = 0$, $m = 1, 2, \cdots, d(\tilde{S})$.

The payoff functions $t\tilde{v}_\lambda(\tilde{S}) = [tv^-_\lambda(\tilde{S}), tv^+_\lambda(\tilde{S})]$ indicates the expected payoffs of the intuitionistic coalition $\tilde{S}$, which are mappings about the coalition $\tilde{S}$ to the interval value set $\bar{R}$, namely $t\tilde{v} : \tilde{F}(N) \to \bar{R}$, and $t\tilde{v}(\varnothing) = 0$.

3 AT Solution for Cooperative Game with Intuitionistic Fuzzy Coalition

3.1 The Mean Value of Preference

In 1988, Yager proposed an ordered weighted average operator (OWA for short) to weighted average of a set of discrete real numbers after ordering. Then Yager [21] developed a continuous ordered weighted arithmetic averaging (C-OWA) operator to aggregate all the numbers within an interval. Nevertheless, the operators only focus the ordered position aggregation over a closed interval $[a, b]$, the aggregation of given importance of $[a, b]$ is not taken into account. In order to integrate the fuzzy data with preference information in decision-making, Lin and Jiang [22] defined a continuous mixed aggregator based on the weight of data and the preference attitude of decision-makers.

A continuous hybrid weighted quasi-arithmetical averaging operator is a mapping $g : \Omega \to R$, which is associated with a Basic Unit-intervals Monotonic (BUM for short) function: $Q : [0,1] \to [0,1]$ satisfies the following three conditions: (i) $Q(0) = 0$; (ii) $Q(1) = 1$; (iii) For arbitrary $a \in [0,1]$, and $b \in [0,1]$, if $a > b$, then $Q(a) \geq Q(b)$, such as:

$$F_Q([a,b]) = h^{-1}\left(\frac{\int_0^1 \alpha(t)\frac{dQ(x)}{dx}h(t)dx}{\int_0^1 \alpha(t)\frac{dQ(x)}{dx}dx}\right) \tag{5}$$

where $t = h^{-1}(h(b) - (h(b) - h(a))x)$, $\alpha(t) \geq 0 (t \in [a,b])$ is a continuous weighting function, Ω is the set of closed intervals, h is a continuous strictly monotonic function, h^{-} is the inverse function of h.

For simplicity, we suppose that the importance of given data is same, $\alpha(t) = 1/2$, $h(t) = t$ in Eq. (5), and there is any BUM function $Q(x) \in \Gamma$, then we can get mean value of the risk preference

$$\begin{aligned} F_Q([a,b]) &= \frac{\int_0^1 \frac{dQ(x)}{dx}[b-(b-a)x]dx}{Q(1)-Q(0)} \\ &= \int_0^1 [b-(b-a)x]dQ(x) \\ &= [b-(b-a)x]Q(x)|_0^1 + (b-a)\int_0^1 Q(x)dx \\ &= a[1-\int_0^1 Q(x)dx] + b\int_0^1 Q(x)dx \\ &= a(1-\theta) + b\theta \end{aligned} \tag{6}$$

It is obviously, the aggregation of $[a, b]$ is same as Yager's C-OWA operator when the given data have same important. And when the importance of data is different, we can use other $\alpha(t)$ functions to weighted aggregate the continuous hybrid information.

It is easily seen from Eq. (6) that the mean value of risk preference $F_Q([a,b])$ is a combination of a and b, where the functions $1-\int_0^1 Q(y)dy$ and $\int_0^1 Q(y)dy$ are the risk attitude coefficients of the fuzzy number $\tilde{A}$ correlating a BUM function $Q(y)$. Yager has proved that the bigger the value of θ, the higher the degree of player's risk preference. We can treat the risk attitude coefficient as the degree of optimism.

3.2 The Characteristic Function with Intuitionistic Fuzzy Coalition Based upon Preference Form

Using the BUM function Q_i and the attitude factor $\theta_i = \int_0^1 Q_i(y)\mathrm{d}y$ of a player i, an attitude factor function $\theta_{\tilde{S}}$ of a fuzzy coalition $\tilde{S}$ in $(N, \tilde{v}, L)$ could be defined as follows.

Definition 7. Let $(N, \tilde{v}, L)$ be a graph cooperative game with intuitionistic fuzzy coalition. For any intuitionistic fuzzy coalition $\tilde{S} \in F(N)$, the risk preference factor $\theta_{\tilde{S}}$ of coalition $\tilde{S}$ is $\arg\min_{\theta}\{\sum_{i \in \mathrm{Supp}(\tilde{S})} (\theta - \theta_i)^2\}$.

Solving the objective function $\arg\min_{\theta}\{\sum_{i \in \mathrm{Supp}(\tilde{S})} (\theta - \theta_i)^2\}$, produces the following equation:

$$\theta_{\tilde{S}} = \frac{1}{|\tilde{S}|} \sum_{i \in \mathrm{Supp}(\tilde{S})} \theta_i \tag{7}$$

where $|\tilde{S}|$ denotes the number of elements in the support set $\mathrm{Supp}(\tilde{S})$.

Equation (7) shows that the risk preference attitude of a coalition is the arithmetic mean value of that of the players therein, which means that the closer the risk preferences of players in a coalition, the more they are likely to agree with the coalition, then making the potential alliance more stable.

Definition 8. Let $(N, \tilde{v}, L)$ be a graph cooperative game with intuitionistic fuzzy coalition. For any $\tilde{S} \in F(N)$, $\theta_{\tilde{S}}$ indicates the attitude factors of intuitionistic fuzzy coalition $\tilde{S}$. Then the characteristic function of $(N, \tilde{v}, L)$ can be obtained based on the mean value of risk preference:

$$\begin{aligned} \xi_\lambda(\tilde{S}) &= \theta_{\tilde{S}} \sum_{m=1}^{d(\tilde{S})} v([\tilde{S}]_{h^+_{m(\lambda)}})(h^+_{m(\lambda)} - h^+_{m-1(\lambda)}) + (1-\theta_{\tilde{S}}) \sum_{m=1}^{d(\tilde{S})} v([\tilde{S}]_{h^-_{m(\lambda)}})(h^-_{m(\lambda)} - h^-_{m-1(\lambda)}) \\ &= \theta_{\tilde{S}} tv^+_\lambda(\tilde{S}) + (1-\theta_{\tilde{S}}) tv^-_\lambda(\tilde{S}) \end{aligned} \tag{8}$$

The characteristic function of $(N, \tilde{v}, L)$ can be aggregated by the attitude factors of coalitions and the mean value of risk preferences. Therefore, a graph cooperative game with intuitionistic fuzzy coalitions can convert into a crisp graph cooperative game. This method not only describes fuzzy information in cooperative games, but also is helpful for solving allocation functions.

3.3 AT Solution and Its Properties

In this section, we will define the AT solution of graph cooperative games with intuitionistic fuzzy coalition, and discuss some important properties of fuzzy by using the relevant properties of crisp AT solution.

Definition 9. For any a graph cooperative game with intuitionistic fuzzy coalition $(N, \tilde{v}, L)$, $\lambda \in [0, 1]$, the fuzzy AT solution of player i is defined by

$$\tilde{A}T_i(N, \tilde{v}_\lambda, L) = \frac{1}{|B^L|} \sum_{B \in B^L} [\xi_\lambda(B_i) - \sum_{k \in \hat{C}^L(B_i \backslash \{i\})} \xi_\lambda(K)] \quad (i = 1, \ldots, n) \tag{9}$$

Theorem 1. If $(N, \tilde{v}, L)$ is a graph cooperative game with intuitionistic fuzzy coalition, $\lambda \in [0, 1]$, then the fuzzy AT solution $\tilde{A}T(N, \tilde{v}_\lambda, L)$ is an allocation of $(N, \tilde{v}, L)$.

Proof: If $i \notin N$, then $\tilde{A}T_i(N, \tilde{v}, L) = 0$.

If $i \in N$, we have $\sum_{i \in K} \tilde{A}T_i(N, \tilde{v}_\lambda, L) = \xi(K_\lambda)$ for any $K \in \hat{C}^L(N)$ because $\tilde{A}T(N, \tilde{v}_\lambda, L)$ satisfies component efficiency. That is to say, $\tilde{A}T(N, \tilde{v}_\lambda, L)$ satisfies group rationality.

In addition, because

$$\xi_\lambda(B) = \xi_\lambda(\bigcup_{K \in \hat{C}^L(B_i \backslash \{i\})} K \cup \{i\}) \geq \sum_{K \in \hat{C}^L(B_i \backslash \{i\})} \xi_\lambda(K) + \xi_\lambda(\{i\})$$

We obtain

$$\xi_\lambda(B) - \sum_{K \in \hat{C}^L(B_i \backslash \{i\})} \xi_\lambda(K) \geq \xi_\lambda(\{i\})$$

So it follows directly that

$$\tilde{A}T_i(N, \tilde{v}_\lambda, L) = \frac{1}{|B^L|} (\sum_{i \in B \in B^L} \xi_\lambda(\tilde{B}_i) - \sum_{K \in \hat{C}^L(B_i \backslash \{i\})} \xi_\lambda(\tilde{K})) \geq \xi_\lambda(\{i\})$$

Therefore, $\tilde{A}T(N, \tilde{v}_\lambda, L)$ satisfies the group rationality and individual rationality from allocations. Hence, we have proven that an AT solution $\tilde{A}T(N, \tilde{v}_\lambda, L)$ is an allocation of $(N, \tilde{v}, L)$.

Theorem 2. For any graph cooperative game with intuitionistic fuzzy coalitions $(N, \tilde{v}, L)$, there always exists a unique AT solution $\tilde{A}T(N, \tilde{v}_\lambda, L)$ determined by Eq. (9), which satisfies properties of component efficiency, efficiency, additivity and component fairness.

To prove the payoff vector $\tilde{\boldsymbol{AT}}(N,\tilde{v}_\lambda,L) = (\tilde{AT}_1(N,\tilde{v}_\lambda,L),\tilde{AT}_2(N,\tilde{v}_\lambda,L),\cdots,$ $\tilde{AT}_n(N,\tilde{v}_\lambda,L))$ is a unique fuzzy AT solution for $(N,\tilde{v},L)$, we need to prove Eq. (9) satisfies the same properties as crisp AT solution [7].

(1) **Component Efficiency**: According to Eq. (8), $\lambda \in [0,1]$, the characteristic function of a $(N,\tilde{v},L)$ can aggregate into a crisp function. Combining with the component efficiency of the AT solution for a crisp graph cooperative game $\sum_{i\in K} AT_i(N,v,L) = v(K)$, then the fuzzy AT solution of $(N,\tilde{v}_\lambda,L)$ have

$$\sum_{i\in K} \tilde{AT}_i(N,\tilde{v}_\lambda,L) = \xi_\lambda(K)$$

That is to say, $\tilde{AT}(N,\tilde{v}_\lambda,L)$ satisfies component efficiency for any $K \in \hat{C}^L(N)$.

(2) **Efficiency**: According to the component efficiency of the AT solution for a crisp graph cooperative game, if the graph cooperative game is a complete game, the crisp AT solution satisfies the efficiency $\sum_i^n \tilde{AT}_i(N,v,L) = v(N)$. Then, based on Eq. (8), we have

$$\sum_i^n \tilde{AT}_i(N,\tilde{v}_\lambda,L) = \xi_\lambda(N)$$

which satisfies the efficiency.

(3) **Component Fairness**: Because of AT solution $\tilde{AT}(N,\tilde{v}_\lambda,L)$ possesses component efficiency, $\lambda \in [0,1]$, any edge $L\{h,l\}$ in $L(K)$, it holds that

$$\sum_{i\in K^h} AT_i((N,\tilde{v}_\lambda,L)\backslash L\{h,l\}) = \xi_\lambda(K^h),\ \sum_{i\in K^l} AT_i((N,\tilde{v}_\lambda,L)\backslash L\{h,l\}) = \xi_\lambda(K^l)$$

where K^h and K^l express the subset of K of players connected to h and l in the graph that results after deleting the link $L\{h,l\}$.
Combining with the component fairness of crisp AT solution, it directly follows that

$$\frac{1}{|K^h|}\sum_{i\in K^h}\left(\tilde{AT}_i(N,\tilde{v}_\lambda,L) - \sum_{i\in K^h}\tilde{AT}_i((N,\tilde{v}_\lambda,L)\backslash L\{h,l\})\right)$$
$$= \frac{1}{|K^l|}\sum_{i\in K^l}\left(\tilde{AT}_i(N,\tilde{v}_\lambda,L) - \sum_{i\in K^l}\tilde{AT}_i((N,\tilde{v}_\lambda,L)\backslash L\{h,l\})\right)$$

(4) **Additivity**: For any graph cooperative games with intuitionistic fuzzy coalitions, $(N,\tilde{v}_1,L) \in G_{IF}(N)$ and $(N,\tilde{v}_2,L) \in G_{IF}(N)$, $\lambda \in [0,1]$, the linearity of characteristic function of Eq. (9) implies that

$$\begin{aligned}
&\tilde{A}T_i(N,(\tilde{v}_1)_\lambda+(\tilde{v}_2)_\lambda,L)\\
&=\frac{1}{|B^L|}\sum_{B\in B^L}[(\xi_1+\xi_2)_\lambda(B_i)-\sum_{K\in\hat{C}^L(B_i\setminus\{i\})}(\xi_1+\xi_2)_\lambda(K)]\\
&=\frac{1}{|B^L|}\sum_{B\in B^L}[((\xi_1)_\lambda(B_i)+(\xi_2)_\lambda(B_i))-\sum_{K\in\hat{C}^L(B_i\setminus\{i\})}((\xi_1)_\lambda(K)+(\xi_2)_\lambda(K))]\\
&=\frac{1}{|B^L|}\sum_{B\in B^L}\left[\left((\xi_1)_\lambda(B_i)-\sum_{K\in\hat{C}^L(B_i\setminus\{i\})}(\xi_1)_\lambda(K)\right)+\left((\xi_2)_\lambda(B_i)-\sum_{K\in\hat{C}^L(B_i\setminus\{i\})}(\xi_2)_\lambda(K)\right)\right]\\
&=\tilde{A}T_i(N,(\tilde{v}_1)_\lambda,L)+\tilde{A}T_i(N,(\tilde{v}_2)_\lambda,L)
\end{aligned}$$

Thus, we have proven that $\tilde{A}T(N,\tilde{v}_\lambda,L)$ is a fuzzy AT solution of the graph cooperative games with intuitionistic fuzzy coalitions $(N,\tilde{v},L)$. ∎

Theorem 3. If $(N,\tilde{v},L)$ is a complete graph cooperative game with intuitionistic fuzzy coalitions, $\lambda\in[0,1]$, then the fuzzy AT solution $\tilde{A}T(N,\tilde{v}_\lambda,L)$ of $(N,\tilde{v},L)$ is equivalent to the fuzzy Shapley $\tilde{\varphi}(N,\tilde{v}_\lambda)$ value.

Proof: According to Eq. (8), $\lambda\in[0,1]$, the characteristic functions of $(N,\tilde{v},L)$ can be aggregated by the attitude factors of coalitions and the mean value of risk preferences, which means the graph cooperative game with intuitionistic fuzzy coalitions $(N,\tilde{v},L)$ can convert into a crisp graph cooperative game (N,ξ_λ,L). In addition, it has been proven that the AT solution is equivalent to the Shapley value when the crisp cooperative game is a complete graph game [7, 9]. Therefore, we can obtain

$$\tilde{A}T(N,\tilde{v}_\lambda,L)=\tilde{\varphi}(N,\tilde{v}_\lambda)$$

which means when the cooperative game has unrestricted communication structure, the fuzzy AT solution $\tilde{A}T(N,\tilde{v}_\lambda,L)$ is the fuzzy Shapley value $\tilde{\varphi}(N,\tilde{v}_\lambda)$. ∎

4 An Application and Comparison Analysis

4.1 Application of a Profit Allocation Problem

In an alliance exploitation of a river, there are upper, middle and lower reaches in the cooperation. For conciseness, called them player 1, player 2, and player 3 respectively. Because the upstream and downstream cannot communicate directly, so the cooperation in the alliance have a communication restriction. In another words, player 1 and player 3 cannot cooperation directly. The essence of this problem is a cooperative game with limited communication structure, and the all potential coalitions are $\{1,2\},\{2,3\},\{1,2,3\}$. If all the players fully take part in the exploitation alliances, that means everyone can invest 100% resource the alliance requires them to, then the profit of coalitions are: $v(\{1\})=30, v(\{2\})=20, v(\{3\})=50, v(\{1,2\})=100$, $v(\{2,3\})=120$, $v(\{1,2,3\})=280$. But due to the limited capacity/resources, or to reduce investment risk, the three players do not fully participate in it. Three players plan the resources they can invest. Palyer 1–3 can invest 30%, 40% and 50% resource at least, while 50%, 40% and 20% must not be invested respectively. It is obviously, the players have some hesitation degree in the coalitions. It follows that ecological exploitation is a graph cooperative game with coalitions. In addition, the players have different risk preferences: player 1 is risk averse and BUM function is $Q_1(y) = y^2$; player 2 is risk neutral and BUM function is $Q_2(y) = y$; player 3 is risk preferring and BUM function is $Q_2(y) = y^{1/2}$. The profit allocation problem is a typical type of graph cooperative game with intuitionistic fuzzy coalitions.

According to the definition of intuitionistic fuzzy set, we know the participation degrees of three players are $\mu_{\tilde{S}}(1) = 0.3$, $\mu_{\tilde{S}}(2) = 0.4$, $\mu_{\tilde{S}}(3) = 0.5$, while the no participation degrees is $\upsilon_{\tilde{S}}(1) = 0.5$, $\upsilon_{\tilde{S}}(2) = 0.4$, $\upsilon_{\tilde{S}}(3) = 0.2$. For player 1, from the intuitionistic fuzzy coalition, the hesitation degree of taking part in the cooperation is $\pi_{\tilde{S}}(1) = 1 - \mu_{\tilde{S}}(1) - \upsilon_{\tilde{S}}(1) = 0.2$. Therefore, the minimum participation degree is $\eta^-_{\tilde{S}}(1) = 0.3$, and the maximum participation is $\eta^+_{\tilde{S}}(1) = \mu_{\tilde{S}}(1) + \pi_{\tilde{S}}(1) = 0.5$. Other intuitionistic fuzzy coalitions in cooperation of players can be similarly explained. And if all the players have different confident level, such as $\lambda = 0$, $\lambda = 0.3$, $\lambda = 0.5$, $\lambda = 0.7$ and $\lambda = 1$, then their fuzzy coalitions can be obtained by Eq. (2).

Table 1. Intuitionistic fuzzy coalitions of players with different confident level

Player	$\tilde{S}$	$h_{\tilde{S}}$				
		$\lambda=0$	$\lambda=0.3$	$\lambda=0.5$	$\lambda=0.7$	$\lambda=1$
1	$\{<1;0.3,0.5>,0,0\}$	0.3	$[0,3,0.36]$	$[0,3,0.4]$	$[0,3,0.44]$	$[0,3,0.5]$
2	$\{0,<2;0.4,0.4>,0\}$	0.4	$[0,4,0.46]$	$[0,4,0.5]$	$[0,4,0.54]$	$[0,4,0.6]$
3	$\{0,0,<3;0.5,0.2>\}$	0.5	$[0,5,0.59]$	$[0,5,0.65]$	$[0,5,0.71]$	$[0,5,0.8]$

It is clear from Table 1 that hesitation has a significant impact on intuitionistic fuzzy coalitions, and the higher confident levels of players, the bigger participation degrees in fuzzy coalitions.

According to Eq. (4) and $\lambda = 1$, the fuzzy payoff of coalition $\{1,2,3\}$ is

$$\begin{aligned} tv^-_{\lambda=1}(\{1,2,3\}) &= \sum_{m=1}^{d(3)} v([\{1,2,3\}]_{h^-_{m(\lambda=1)}})(h^-_{m(\lambda=1)} - h^-_{m-1(\lambda=1)}) \\ &= v(\{1,2,3\})(h^-_1 - h_0) + v(\{2,3\})(h^-_2 - h^-_1) + v(\{3\})(h^-_3 - h^-_2) \\ &= 280 \times (0.3 - 0) + 120 \times (0.4 - 0.3) + 50 \times (0.5 - 0.4) = 101 \end{aligned}$$

$$\begin{aligned} tv^+_{\lambda=1}(\{1,2,3\}) &= \sum_{m=1}^{d(3)} v([\{1,2,3\}]_{h^+_{m(\lambda=1)}})(h^+_{m(\lambda=1)} - h^+_{m-1(\lambda=1)}) \\ &= v(\{1,2,3\})(h^+_1 - h_0) + v(\{2,3\})(h^+_2 - h^+_1) + v(\{3\})(h^+_3 - h^+_2) \\ &= 280 \times (0.5 - 0) + 120 \times (0.6 - 0.5) + 50 \times (0.8 - 0.6) = 162 \end{aligned}$$

Next, using the BUM functions of three players and Eq. (7), the risk preference factor $\theta_{\{1,2,3\}}$ for the coalition $\{1,2,3\}$ is

$$\theta_{\{1,2,3\}} = \frac{1}{3}\left(\int_0^1 Q_{\{1\}}(y)dy + \int_0^1 Q_{\{2\}}(y)dy + \int_0^1 Q_{\{3\}}(y)dy\right) = \frac{1}{3}\left(\frac{1}{3} + \frac{1}{2} + \frac{2}{3}\right) = \frac{1}{2}$$

Then, using Eq. (8), we can calculate the characteristic function of the coalition $\{1,2,3\}$

$$\begin{aligned}\xi_{\lambda=1}(\{1,2,3\}) &= \theta_{\{1,2,3\}} tv^{+}_{\lambda=1}(\{1,2,3\}) + (1-\theta_{\{1,2,3\}}) tv^{-}_{\lambda=1}(\{1,2,3\}) \\ &= 101 \times \frac{1}{2} + 162 \times \frac{1}{2} = 131.5\end{aligned}$$

In similar way, the risk preference factors and characteristic functions of coalitions can be calculated. Finally, using Eq. (9), we can obtain all the allocation values of players, depicted in Table 2.

Table 2. Allocation values of players in intuitionistic fuzzy coalitions

$IF(N)$	$\theta_{\tilde{S}}$	$tv_{\lambda=1}(\tilde{S})$	$\xi_{\lambda=1}(\tilde{S})$	$\tilde{AT}_1(N,\tilde{v}_{\lambda=1},L)$	$\tilde{AT}_2(N,\tilde{v}_{\lambda=1},L)$	$\tilde{AT}_3(N,\tilde{v}_{\lambda=1},L)$
$\tilde{S}_{\{1\}} = \{<1;0.3,0.5>,0,0\}$	1/3	[9, 15]	11	10	0	0
$\tilde{S}_{\{2\}} = \{0,<2;0.4,0.4>,0\}$	1/2	[8, 12]	10	0	11	0
$\tilde{S}_{\{3\}} = \{0,0,<3;0.5,0.2>\}$	2/3	[25, 40]	35	0	0	35
$\tilde{S}_{\{1,2\}} = \{<1;0.3,0.5>, <2;0.4,0.4>,0\}$	5/12	[32, 52]	40.33	20.665	19.665	0
$\tilde{S}_{\{2,3\}} = \{0,<2;0.4,0.4>, <3;0.5,0.2>\}$	7/12	[53, 82]	69.92	0	22.46	47.46
$\tilde{S}_{\{1,2,3\}} = \{<1;0.3,0.5>, <2;0.4,0.4>, <3;0.5,0.2>\}$	1/2	[101, 162]	131.5	27.86	49.92	53.72

It is easily seen from Table 2 that it is a best choice for all three players to form a grand coalition $\{1,2,3\}$, and everyone can be allocated their maximum returns, which means cooperation is always better than doing alone.

4.2 Comparison and Discussion

(1) Allocation values of players under different confidence levels and risk preferences

For the confident level $\lambda \in [0,1]$, we can obtain the allocation values of player 1, player 2, and player 3 respectively, depicted as in Table 3.

In addition, if we suppose that the players change their risk preferences: player 1 and 2 prefer bigger risk averse, their BUM function become into $Q_1(y) = y^3$ and $Q_2(y) = y^2$; player 3 likes smaller risk preferring and his BUM function becomes into $Q_2(y) = y^{1/3}$.

Table 3. Allocation values of players under different confidence levels

λ	$\tilde{AT}_1(N, \tilde{v}_\lambda, L)$	$\tilde{AT}_2(N, \tilde{v}_\lambda, L)$	$\tilde{AT}_3(N, \tilde{v}_\lambda, L)$
1	27.86	49.92	53.72
0.7	26.10	49.30	46.95
0.5	24,93	44,63	46.69
0.3	23.76	42.51	43.88
0	22	39.33	39.67

Table 4. Allocation values of players under different risk preferences

$IF(N)$	$\lambda_{\tilde{S}}$	$tv_{\lambda=1}(\tilde{S})$	$\xi_{\lambda=1}(\tilde{S})$	$\tilde{AT}_1(N, \tilde{v}_{\lambda=1}, L)$	$\tilde{AT}_2(N, \tilde{v}_{\lambda=1}, L)$	$\tilde{AT}_3(N, \tilde{v}_{\lambda=1}, L)$
$\tilde{S}_{\{1\}}\tilde{S}_{\{1\}}$	1/4	[9, 15]	10.5	10.5	0	0
$\tilde{S}_{\{2\}}$	1/3	[8, 12]	9.33	0	9.33	0
$\tilde{S}_{\{3\}}$	3/4	[25, 40]	36.25	0	0	36.25
$\tilde{S}_{\{1,2\}}$	7/24	[32, 52]	37.83	19.5	18.33	0
$\tilde{S}_{\{2,3\}}$	13/24	[53, 82]	68.71	0	20.895	47.815
$\tilde{S}_{\{1,2,3\}}$	4/9	[101, 162]	128.11	26.8	47.05	54.26

It can be easily seen that from Tables 2, 3 and 4 that the allocation values of all the players directly depends on their own confidence levels/preference regardless of other player's. And it is showed a positive correlation. That is to say, the bigger a player's confidence levels/risk preference, the bigger the player's expected pay off value.

(2) Allocation values of players by Shapley value

In addition, as we all known, the best allocation value of cooperative games is Shapley value. If we can suppose that player 1 and player 3 can cooperation directly in cooperation, which means the cooperative game with complete communication structure, we can compare the fuzzy AT solution with the fuzzy Shapley value. In this situation, the coalition payoff $\tilde{v}(\{1,3\})$ is the sum of payoffs of player 1 and player 3, $\tilde{v}(\{1,3\}) = \tilde{v}(\{1\}) + \tilde{v}(\{3\}) = 80$ in original allocation problem.

Then, using Eqs. (6)–(8), we can obtain the risk preference factor and the characteristic function of the coalition $\{1,3\}$, and the allocation values of players by the Shapley value listed in Table 3.

From Tables 2 and 5, we can obtain $AT_1(N, \tilde{v}_{\lambda=1}, L) < \varphi_1(N, \tilde{v}_{\lambda=1})$, $AT_2(N, \tilde{v}_{\lambda=1}, L) > \varphi_2(N, \tilde{v}_{\lambda=1})$ and $AT_3(N, \tilde{v}_{\lambda=1}, L) < \varphi_3(N, \tilde{v}_{\lambda=1})$. That is to say, if all the players have the same confidence levels $\lambda = 1$ for the hesitation degrees when they participate to coalitions, the allocation of player 2 increases while player 1 and player 3 both decrease according to the fuzzy AT solution, relative to the fuzzy Shapley value. This is a result of highlighting the special status of player 2, which is critical player in the cooperative games. It shows that the players' profitability depend not only on their marginal contribution degree to the coalition, but also on the communication structure of the coalition and players' positions in the cooperative game.

Table 5. Allocation of players in intuitionistic fuzzy coalitions by Shapley values

$IF(N)$	$\lambda_{\tilde{S}}$	$tv_{\lambda=1}(\tilde{S})$	$\xi_{\lambda=1}(\tilde{S})$	$\tilde{\varphi}_1(N,\tilde{v}_{\lambda=1})$	$\tilde{\varphi}_2(N,\tilde{v}_{\lambda=1})$	$\tilde{\varphi}_3(N,\tilde{v}_{\lambda=1})$
$\tilde{S}_{\{1\}}$	1/3	[9, 15]	11	11	0	0
$\tilde{S}_{\{2\}}$	1/2	[8, 12]	10	0	10	0
$\tilde{S}_{\{3\}}$	2/3	[25, 40]	35	0	0	35
$\tilde{S}_{\{1,2\}}$	5/12	[32, 52]	40.33	20.665	19.665	0
$\tilde{S}_{\{2,3\}}$	7/12	[53, 82]	69.92	0	22.46	47.46
$\tilde{S}_{\{1,3\}}$	1/2	[34, 55]	44.5	10.25	0	34.25
$\tilde{S}_{\{1,2,3\}}$	1/2	[101, 162]	131.5	30.83	43.04	57.63

5 Conclusion

Cooperative games with limited communication structure and fuzzy information are more applicable to depict cooperation actions. Engaging in cooperation and estimating profits mainly depends on players' judgments and intuition, especially the degree of participation in alliances. The degrees are often vague and not easy to be represented with crisp values and fuzzy numbers, there are some hesitations in practical cooperation. This paper defines a graph cooperative with intuitionistic fuzzy coalition remarkably differ from fuzzy coalition, then we propose an AT solution with confidence levels based on the defined preference weighted form, which is a general case of the classical cooperative game and cooperative game with fuzzy coalitions. The method and model proposed in this paper can depict the restrictiveness and fuzziness of real alliances, and it is helpful for solving allocation problems. This study extends the fuzzy cooperative game theory and integrates different risk preferences of players into cooperation. In addition, the Shapley value is a special form of the AT solution, this study shows the players' profitability both depend on their marginal contribution degree to the coalition and the communication structure of the coalition.

Acknowledgments. This research was partially supported by the Natural Science Foundation of China (Nos. 71572040, 71601049), the Science Foundation of Ministry of Education of China (No. 19YJC630201), the Project for Ecological Civilization Research of Fujian Social Science Research Base (Nos. KXJD1813A, KXJD1837A), the Program for Distinguished Young Scholars in University of Fujian Province (No. K80SCC55A), and the Project of Fujian Agriculture and Forestry University (No. XJQ201635). We appreciate the comments and suggestions will give by the reviewers and editor of this journal.

References

1. Myerson, R.B.: Graphs and cooperation in games. Math. Oper. Res. **2**, 225–229 (1977)
2. Meng, F., Tian, D.: The Banzhaf value for fuzzy games with a coalition structure. Res. J. Appl. Sci. Eng. Technol. **14**(1), 22–34 (2012)

3. Fernández, J.R., Gallego, I., Jiménez-Losada, A., Ordóñez, M.: Cooperation among agents with a proximity relation. Eur. J. Oper. Res. **250**(2), 555–565 (2016)
4. Li, X., Sun, H., Hou, D.: On the position value for communication situations with fuzzy coalition. J. Intell. Fuzzy Syst. **33**(1), 113–124 (2017)
5. Herings, P.J.J., Van Der Laan, G., Talman, D.: The average tree solution for cycle-free graph games. Games Econ. Behav. **62**(1), 77–92 (2008)
6. Talman, D., Yamamoto, Y.: Average tree solution and subcore for acyclic graph games. J. Oper. Res. Soc. Jpn. **51**(3), 203–212 (2008)
7. Herings, P.J.J., van der Laan, G., Talman, A.J.J., Yang, Z.: The average tree solution for cooperative games with communication structure. Games Econ. Behav. **68**(2), 626–633 (2010)
8. Van den Brink, R., Herings, P.J.J., van der Laan, G., Talman, D.J.J.: The average tree permission value for games with a permission tree. Econ. Theory **58**(1), 99–123 (2013)
9. Mishra, D., Talman, A.J.J.: A characterization of the average tree solution for tree games. Int. J. Game Theory **39**(1–2), 105–111 (2010)
10. Aubin, J.P.: mathematical methods of game and economic theory. Stud. Math. Appl. **235**(1), 19–30 (1982)
11. Jiménez-Losada, A., Fernández, J.R., Ordóñez, M.: Myerson values for games with fuzzy communication structure. Fuzzy Sets Syst. **213**(3), 74–90 (2013)
12. Xu, G., Li, X., Sun, H., Su, J.: The Myerson value for cooperative games on communication structure with fuzzy coalition. J. Intell. Fuzzy Syst. **33**(1), 27–39 (2017)
13. Basallote, M., Hernández-Mancera, C., Jiménez-Losada, A.: A new Shapley value for games with fuzzy coalitions. Fuzzy Sets Syst. (2019). https://doi.org/10.1016/j.fss.2018.12.018
14. Atanassov, K.T.: Intuitionistic fuzzy sets. Fuzzy Sets Syst. **20**(1), 87–96 (1986)
15. Li, D.F.: Decision and Game Theory in Management with Intuitionistic Fuzzy Sets. Springer, Heidelberg (2014). https://doi.org/10.1007/978-3-642-40712-3
16. Fodor, J.C., Roubens, M.R.: Fuzzy Preference Modelling and Multicriteria Decision Support. Springer, Dordrecht (2013)
17. Li, K.W., Inohara, T., Xu, H.: Coalition analysis with preference uncertainty in group decision support. Appl. Math. Comput. **231**, 307–319 (2014)
18. Lin, J., Meng, F., Chen, R.Q., Zhang, Q.: Preference attitude-based method for ranking intuitionistic fuzzy numbers and its application in renewable energy selection. Complexity **1**, 1–14 (2018)
19. Murofushi, T., Sugeno, M.: An interpretation of fuzzy measures and the Choquet integral as an integral with respect to a fuzzy measure. Fuzzy Sets Syst. **29**(2), 201–227 (1989)
20. Tsurumi, M., Tanino, T., Inuiguchi, M.: A Shapley function on a class of cooperative fuzzy games. Eur. J. Oper. Res. **129**(3), 596–618 (2001)
21. Yager, R.R.: OWA aggregation over a continuous interval argument with applications to decision making. IEEE Trans. Syst. Man Cybern. Part B **34**(5), 1952–1963 (2004)
22. Lin, J., Jiang, Y.: Some hybrid weighted averaging operators and their application to decision making. Inf. Fusion **16**(1), 18–28 (2014)

Shapley Value Method for Benefit Distribution of Technology Innovation in Construction Industry with Intuitionistic Fuzzy Coalition

Ting Han(✉)

Fuzhou University of International Studies and Trade,
Fuzhou, People's Republic of China
272993345@qq.com

Abstract. The formation of the technology innovation coalition of the construction industry can give full play to the resource advantages of all participants, innovate technologies, save cost, improve construction quality, and achieve a multi-win situation. The key to the success of the coalition is to establish a fair and efficient mechanism of benefit distribution. Firstly, the forming mechanism and value creation mechanism is analyzed. Then the benefit distribution under the condition that members have certain degree of participation and certain degree of non-participation in the coalition is discussed, assuming that the members are fully aware of the expected benefit of different cooperation strategies before the cooperation. The essence is to solve cooperative game with intuitionistic fuzzy coalition. In this paper, Shapley value for intuitionistic fuzzy cooperative game is proposed by taking use of intuitionistic fuzzy set theory, Choquet integrals and continuous ordered weighted average operator. It's also proofed that the defined Shapley value satisfies three axioms. Finally, the effectiveness and rationality of Shapley is illustrated by a numerical example.

Keywords: Intuitionistic · Fuzzy coalition · Technology innovation in construction industry · Benefit distribution · Shapley value method

1 Introduction

The construction industry is the pillar industry of our national economy. With the rapid development of urbanization, the construction industry is still in the growth stage, and its growth and development speed depends on the innovation of the construction industry. The improvement of the innovation ability of construction companies can enhance their economic viability and is also the main source of competitive advantage. The innovation content of construction industry mainly includes concept innovation, system innovation, scientific and technological innovation [1]. Among them, the improvement of technological innovation which is the most direct innovation content is of vital importance to the development of national economy. It can solve the difficult,

Supported by the education and scientific research project of young and middle-aged teachers in Fujian province (JAT170728).

D. Li (Ed.): EAGT 2019, CCIS 1082, pp. 85–104, 2019.
https://doi.org/10.1007/978-981-15-0657-4_6

new and large-scale projects that can not be completed or can be completed by technology before, but can not be economically completed.

Technological innovation costs. It is very difficult for construction enterprises, relevant universities or research institutes, or government departments to carry out technological innovation independently, and it will lead to the disconnection between theory and practice and waste of resources. Inter-firm coalitions can promote technological innovation by integrating internal and external resources [2]. Technological innovation coalitions of construction industry emerge as the times require. It can give full play to the resource advantages of all participants, combine theory and practice, innovate technology, save resources and improve construction quality. Whether a coalition can be formed and the efficiency and stability of its operation after its formation largely depend on whether it can distribute benefit fairly and efficiently.

2 Research Status at Home and Abroad

2.1 Research Status of Technical Innovation in Construction Industry

Kangari [3] believes that architectural technology innovation is increasingly becoming an important factor affecting the development of the construction industry, and summarizes the four major factors affecting the development of Japanese architecture: (1) strategic alliance; (2) access to effective information; (3) reputation for innovation; (4) technological innovation. Arditi [4] summarized the experience in the innovation and development of construction equipment in the last three decades of the last century, and the impetus for industrial innovation is the stimulation of technological progress and the driving role of the market. Finally, it is concluded that continuous technological innovation is the catalyst for the development of the industry. Pries and Janszen [5] analyzed the innovation and strategic behaviors of the construction industry and believed that the external environment has a great impact on innovation. When the environment plays a positive role, the speed of industrial innovation will be intensified. When the environment is unfavorable, a relatively high level of management is required. Eriksson [6] studied the way of project delivery, which showed that the possible choice of project delivery method played an important role in the level of trust between project participants. Horta [7] studied the development trend of Portuguese construction industry, and used the principle of "assuming doubtful points to be true" to strengthen the construction of composite indicators. The Portuguese construction industry experienced a significant performance improvement in the 1990s, but this growth trend has slowed in recent years. The conclusion is that strong corporate performance is influenced by the national economic background, and small professional companies and large contractors tend to achieve the best performance level.

Šuman [8] believes that the improvement of the innovation ability of construction companies can enhance their economic viability. Although innovation is a risky behavior with a low probability of success, it is impossible to survive without innovation. Blayse and Manley [9] summarized six main influencing factors that promote or hinder innovation in the construction industry: (1) customers and manufacturers; (2) industrial structure; (3) the relationship between individuals and enterprises, as well as between internal and external parties in the industry; (4) procurement system; (5) regulations/standards; (6) the nature and quality of organizational resources. Fruin et al. [2] believe that inter-company alliances help companies acquire and integrate internal and external resources to promote technological innovation. In particular, it analyzes the alliance governance structure and governance mechanisms, and shows how to protect and improve web-based innovation and competitive advantage within a decade.

In short, there are still many theoretical researches on the technical innovation of the construction industry at home and abroad, which provides some theoretical guidance for the further technical innovation of the construction industry in China. However, the current literature mainly focuses on the effect of management means on the formation of alliance, while the research on the formation mechanism of alliance is not practical enough.

2.2 Research Status of Benefit Distribution of Coalition

Jahn et al. [10] studied the profit distribution of capacity units under the non-hierarchical production cooperation network. Every enterprise that could play a role in the value chain was called a capacity unit, and the evaluation model of capacity unit under the condition of information asymmetry was proposed. Chauhan et al. [11] studied the supply chain partnership based on benefit sharing, proposed the benefit distribution model of the retail two-level supply chain system, and distributed the total benefit according to the risk proportion of the members in the supply chain. Canakoglu et al. [12] studied the two-stage wireless communication supply chain system with technology-independent demand, determined the benefit distribution strategy based on the proportion of technology investment, and obtained Nash equilibrium of benefit distribution by using non-cooperative games. Jia and Yokoyama [13] put forward the benefit distribution strategy of joint sales of small manufacturers that have independent viability in the electronic product market by making use of the cooperative strategy. Sakawa et al. [14] proposed a fuzzy planning model for production and transportation coordination, and determined the profit distribution strategy between production and transportation sectors by using the solution concept and reconciliation method in cooperative games.

From the perspective of cost, Zheng et al. [15] investigated the contribution of enterprises to the alliance according to the principal-agent theory. If an enterprise wants to obtain the profit distribution consistent with its contribution, it must satisfy the enterprise's efforts to contribute to the cost of innovation. Feng and Chen [16] proposed the method of distributing the total interests according to the proportion of interests, and determined the proportion of interests by using the fuzzy comprehensive evaluation method based on the risks taken by partner investment. Dai et al. [17] applied Shapley value method of multiplayer cooperative strategy to distribute benefit distribution of aligned cooperative enterprises, and proposed a correction algorithm based on risk factors. Liao [18] created a profit distribution model on the basis of enterprise alliance, conducted correlation analysis, and proposed a secondary benefit distribution mechanism on this basis, combining the benefit distribution strategy with the incentive strategy and the punishment strategy with the constraint method, so as to make the enterprise alliance reach the Pareto optimal under certain conditions. Sang et al. [19] utilized interval Shapley value method and introduced risk factors and financing costs of alliance enterprises to distribute benefit among enterprises. Han et al. [20] pointed out that their behavior of interest distribution conforms to the applicable conditions of cooperative countermeasures, and constructed a two-person cooperative game model of knowledge creation interest distribution among enterprises. Chen et al. [21] proposed a benefit distribution scheme for supply chain enterprises based on orthogonal projection method. According to the weight coefficients of different distribution schemes for nodal enterprises in the supply chain, the results obtained by various distribution methods are compromised into a comprehensive benefit distribution scheme. Chen and Zhang [22] defined the payment function and the Shapley value of fuzzy alliance cooperative game by Choquet integral of fuzzy measure, and proposed the benefit distribution strategy of enterprise alliance based on fuzzy alliance cooperative game. Tan [23] defined the existence of Shapley value of n-person game with interval fuzzy coalition, and gave an interpretation expression of Shapley value of such game, which was applied to the benefit distribution of supply chain cooperative enterprises.

It is not difficult to find that the current research mainly focuses on how to distribute the benefits when participants participate fully or partially in the alliance, while ignoring how to distribute the benefits when participants participate to some extent in the alliance and not to some extent.

2.3 The Research Status of Intuitionistic Fuzzy Set Theory and Its Application at Home and Abroad

In 1983, professor Atanassov [24, 25] proposed the concept of intuitionistic fuzzy sets on the basis of introducing another scale, namely degree of non-membership. Since the intuitionistic fuzzy set was proposed, it has attracted worldwide attention and research. In the first 10 years, most of these researches are about the concepts, properties, set operations, logical operators, correlation coefficients, similarity, topology and other basic theories of intuitionistic fuzzy sets, mainly from the mathematical perspective. In the 21st century, on the one hand, more in-depth theoretical research has been carried out on the original basis; on the other hand, relevant application research of intuitionistic fuzzy sets has also been carried out. It includes the application of group decision [26], multi-attribute decision [27], intuitionistic fuzzy set in fuzzy reasoning [28], intuitionistic fuzzy clustering analysis [29] and fault diagnosis [30].

At present, there are few researches on the application of intuitionistic fuzzy sets in game theory. There are great differences in the research methods and contents when using intuitionistic fuzzy set and fuzzy set to study the game problem: the former uses the two-scale intuitionistic fuzzy set to solve the fuzziness in the game problem, and needs to carry out the comparison of the conflicting two-scale size, that is, the comparison between the vectors composed of degree of membership and degree of non-membership; The latter describes the fuzziness with a single scale fuzzy set and compares the size of a single scale, that is, between real Numbers. These differences make it impossible to simply apply the existing fuzzy game theory and methods to solve the problem of intuitionistic fuzzy set games.

In 2009, Li [31] gave the mathematical representation, definition and properties of the matrix game whose payoff value is an intuitionistic fuzzy set. In 2010, the above research work was generalized, and the matrix game solution method with the payment value as the interval value intuitionistic fuzzy set was proposed [32]. Meanwhile, the matrix game theoretical model with the payment value as the triangle intuitionistic fuzzy number and the dictionary order solution method were proposed [33]. Nayak and Pal [34] studied the solving method of bimatrix games with real payoff values and expected targets as intuitionistic fuzzy sets by using the construction method of intuitionistic fuzzy expected targets proposed by Angelov [35].

The above researches focus on the two-person zero sum or non-zero sum non-cooperative game of the intuitionistic fuzzy set with the payoff value or the expected goal (interval value), that is, the matrix game and the bimatrix game, and have not yet involved the concept, property and solution method of n-person cooperative game with intuitionistic fuzzy coalition.

3 Formation and Value Creation Mechanism of Intuitionistic Fuzzy Coalition of Technological Innovation in Construction Industry

3.1 Formation Mechanism of Intuitionistic Fuzzy Coalitions for Technological Innovation in Construction Industry

The technology innovation coalitions of the construction industry refers to a cooperative organization formed by two or three of the four parties, including the relevant government departments, research institutions, relevant universities and construction enterprises, for a certain purpose, with industrial technology innovation as the core, resource sharing as the method, risk sharing and benefit sharing. Each participant of the coalition has its own core resources or competitiveness, including human resources, material resources, financial resources, policy support and procedural work. Among them, government departments can provide corresponding policy support, promote the use of new technologies developed by the coalition, or jointly provide scientific research funds with enterprises, that is, the core resources of government departments are mainly policy support and financial resources. The core resource of scientific research institutions or relevant institutions of higher learning is human resources, which is responsible for proposing and solving innovative solutions. Construction enterprises (including the investigation of the construction units, construction units, design units, materials and equipment supply unit, supervision and consulting, etc.) are the main demanders of technology innovation, They understand the needs of the market, has a strong financial strength and a strong practical ability, with a variety of equipment, materials and venues, Compared with other participants, it has a great advantage in completing the procedural work required in the establishment and operation of the coalition, that is, the core resources of construction enterprises are material resources, financial resources and procedural work. Thus, it solves the disadvantages of enterprises' lack of research and development ability, scientific research institutions and relevant colleges and universities' lack of funds, and governments' lack of both. Resources and advantages can be complemented through the coalition.

After the formation of the coalition, its members will face various controllable and uncontrollable risks, covering market, capital, technology, management, nature and politics. Different coalition members face different types and sizes of risk. In fact, the members of the technology coalition of the construction industry are more likely to participate in the coalition to a certain extent and not to participate in the coalition to a certain extent. The coalition members are somewhat hesitant to participate in the coalition, that is, the formed technology innovation coalition is the intuitionistic fuzzy coalition.

3.2 The Mathematical Description of the Intuitionistic Fuzzy Coalitions of Technology Innovation in Construction Industry

The Concept and Representation of Intuitionistic Fuzzy Sets

In 1983, professor Atanassov [24, 25] proposed the concept of intuitionistic fuzzy sets on the basis of introducing another scale, namely degree of non-membership.

Definition 1. [36] Let X be a domain in which x is any given element. If two mappings $\mu_{\tilde{A}} : X \to [0,\ 1]$ and $\upsilon_{\tilde{A}} : X \to [0, 1]$ make $x \in X \to \mu_{\tilde{A}}(x) \in [0, 1], \upsilon_{\tilde{A}}(x) \in [0,\ 1]$ and $0 \leq \mu_{\tilde{A}}(x) + \upsilon_{\tilde{A}}(x) \leq 1$, we say they determine one intuitionistic fuzzy set in the field of X, which can be abbreviated as $\tilde{A} = \{ < x, \mu_{\tilde{A}}(x), \upsilon_{\tilde{A}}(x) > | x \in X \}$, $\mu_{\tilde{A}}$ and $\upsilon_{\tilde{A}}$ are called the membership function and the non-membership function respectively. The set composed of all intuitionistic fuzzy sets on the theoretical domain is denoted as $G_{if}(X)$.

It can be seen from Definition 1 that the degree of membership and degree of non-membership of the intuitionistic fuzzy set are almost independent of each other, and the only requirement is that the sum of the two is no more than 1.

When the domain is finite, namely, $X = \{x_1, x_2, \cdots, x_n\}$, the intuitionistic fuzzy set $\tilde{A}$ can be expressed as follows.

1. $\tilde{A} = < x_1, \mu_{\tilde{A}}(x_1), \upsilon_{\tilde{A}}(x_1) > + < x_2, \mu_{\tilde{A}}(x_2), \upsilon_{\tilde{A}}(x_2) > + \cdots + < x_n, \mu_{\tilde{A}}(x_n), \upsilon_{\tilde{A}}(x_n) >$
$$= \sum_{j=1}^{n} < x_j, \mu_{\tilde{A}}(x_j), \upsilon_{\tilde{A}}(x_j) >$$
2. $\tilde{A} = < \mu_{\tilde{A}}(x_1), \upsilon_{\tilde{A}}(x_1) > /x_1 + < \mu_{\tilde{A}}(x_2), \upsilon_{\tilde{A}}(x_2) > /x_2 + \cdots + < \mu_{\tilde{A}}(x_n), \upsilon_{\tilde{A}}(x_n) > /x_n$
$$= \sum_{j=1}^{n} < \mu_{\tilde{A}}(x_j), \upsilon_{\tilde{A}}(x_j) > /x_j$$
3. $$\tilde{A} = \frac{< \mu_{\tilde{A}}(x_1), \upsilon_{\tilde{A}}(x_1) >}{x_1} + \frac{< \mu_{\tilde{A}}(x_2), \upsilon_{\tilde{A}}(x_2) >}{x_2} + \cdots + \frac{< \mu_{\tilde{A}}(x_n), \upsilon_{\tilde{A}}(x_n) >}{x_n}$$
$$= \sum_{j=1}^{n} \frac{< \mu_{\tilde{A}}(x_j), \upsilon_{\tilde{A}}(x_j) >}{x_j}$$
4. $\tilde{A} = \left(< \mu_{\tilde{A}}(x_1), \upsilon_{\tilde{A}}(x_1) > , < \mu_{\tilde{A}}(x_2), \upsilon_{\tilde{A}}(x_2) > , \cdots, < \mu_{\tilde{A}}(x_n), \upsilon_{\tilde{A}}(x_n) > \right)$

In the formula, "+" or "$\sum$" do not represent ordinary addition and sum; Neither "/" nor "–" has the meaning of a fraction; The term of 0 for both degree of membership and degree of non-membership is generally not written.

When the domain X is infinitely discrete or continuous, the intuitionistic fuzzy set $\tilde{A}$ can be expressed as:

1. $\tilde{A} = \int_{x \in X} <x, \mu_{\tilde{A}}(x), \upsilon_{\tilde{A}}(x)>$
2. $\tilde{A} = \int_{x \in X} <\mu_{\tilde{A}}(x), \upsilon_{\tilde{A}}(x)> /x$
3. $\tilde{A} = \int_{x \in X} \frac{<\mu_{\tilde{A}}(x), \upsilon_{\tilde{A}}(x)>}{x}$

where, "$\int$" does not have the meaning of integral, but represents all the elements $x \in X$ and their degrees of membership and non-membership.

In this paper, union and intersection of two intuitionistic fuzzy sets are defined as:

$$\tilde{A}U\tilde{B} = \{x, \mu_{\tilde{A}U\tilde{B}}(x), \upsilon_{\tilde{A}U\tilde{B}}(x) | x \in X\}$$

$$\tilde{A} \cap \tilde{B} = \{x, \mu_{\tilde{A}\cap\tilde{B}}(x), \upsilon_{\tilde{A}\cap\tilde{B}}(x) | x \in X\}$$

where

$$\mu_{(\tilde{A}\cup\tilde{B})}(x) = \mu_{\tilde{A}}(x) \vee \mu_{\tilde{B}}(x)$$

$$\upsilon_{(\tilde{A}\cup\tilde{B})}(x) = \upsilon_{\tilde{A}}(x) \wedge \upsilon_{\tilde{B}}(x)$$

$$\mu_{(\tilde{A}\cap\tilde{B})}(x) = \mu_{\tilde{A}}(x) \wedge \mu_{\tilde{B}}(x)$$

$$\upsilon_{(\tilde{A}\cap\tilde{B})}(x) = \upsilon_{\tilde{A}}(x) \vee \upsilon_{\tilde{B}}(x)$$

They satisfy the relations as usual, i.e.:

$$(\tilde{A}\cap\tilde{B}) \subseteq \tilde{B} \subseteq (\tilde{A}\cup\tilde{B})$$

The Technology Innovation Intuitionistic Fuzzy Coalition of Construction Industry

According to Definition 1, the degree of membership is used to indicate the degree to which the member participates in the technology innovation coalition, while the degree of non-membership is used to indicate the degree to which the member does not participate in the coalition. The cooperative game with fuzzy coalition (N, σ) is extended to the cooperative game with intuitionistic fuzzy coalition (N, γ). The set composed of all intuitionistic fuzzy sets on the theoretical domain is denoted as $G_{if}(X) \cdot \gamma$ is the payment function of intuitionistic fuzzy coalition, satisfying $\gamma : G_{if}(N) \rightarrow [0, +\infty]$ and $\gamma(\varnothing) = 0$. For any intuitionistic fuzzy coalition, the participation degree of members to the coalition is $\mu_{\tilde{S}}(i) \in [0, 1]$, the non-participation degree is $\upsilon_{\tilde{S}}(i) \in [0, 1]$, and $0 \leq \mu_{\tilde{S}}(x) + \upsilon_{\tilde{S}}(x) \leq 1$. Thus the hesitation degree of members to participate in the coalition is $\pi_{\tilde{S}}(i) = 1 - \mu_{\tilde{S}}(x) - \upsilon_{\tilde{S}}(x)$, when the hesitation degree of all members is zero, the intuitionistic fuzzy coalition degenerates into the fuzzy coalition. Therefore, the fuzzy coalition is a special case of the intuitionistic fuzzy coalition, and the intuitionistic fuzzy coalition is an extension of the fuzzy coalition.

According to the representation method of intuitionistic fuzzy sets, any intuitionistic fuzzy coalition can be expressed as:

$$\tilde{S} = \{ <1, \mu_{\tilde{S}}(1), \upsilon_{\tilde{S}}(1)>, <2, \mu_{\tilde{S}}(2), \upsilon_{\tilde{S}}(2)>, \cdots, <n, \mu_{\tilde{S}}(n), \upsilon_{\tilde{S}}(n)> \}$$

The membership degree of member i to intuitionistic fuzzy coalition $\tilde{S} \in G_{if}(N)$ is $\mu_{\tilde{S}}(i)$, the non-membership degree is $\upsilon_{\tilde{S}}(i)$, The degree of hesitation is $\pi_{\tilde{S}}(i) = 1 - \mu_{\tilde{S}}(x) - \upsilon_{\tilde{S}}(x)$, The distribution of hesitation will change the actual participation and the actual non-participation of the members. When the degree of hesitation is all distributed to the degree of participation, the degree of participation of the members in the intuitionistic fuzzy coalition is the largest, which is $\mu_{\tilde{S}}(i) + \pi_{\tilde{S}}(i)$. Therefore, in the cooperation, members' participation in the intuitionistic fuzzy coalition should be between $\mu_{\tilde{S}}(i)$ and $\mu_{\tilde{S}}(i) + \pi_{\tilde{S}}(i)$, that is, in the closed interval $[\mu_{\tilde{S}}^-(i), \mu_{\tilde{S}}^+(i)] = [\mu_{\tilde{S}}(i), \mu_{\tilde{S}}(i) + \pi_{\tilde{S}}(i)]$, where the lower limit is $\mu_{\tilde{S}}^-(i) = \mu_{\tilde{S}}(i)$ and the upper limit is $\mu_{\tilde{S}}^+(i) = \mu_{\tilde{S}}(i) + \pi_{\tilde{S}}(i) = 1 - \upsilon_{\tilde{S}}(i)$. Obviously, for member i $0 \leq \mu_{\tilde{S}}^-(i) \leq \mu_{\tilde{S}}^+(i) \leq 1$, Moreover, the intuitionistic fuzzy union can be expressed as the interval value $\tilde{S} = \{<1, [\mu_{\tilde{S}}(1), 1 - \upsilon_{\tilde{S}}(1)]>, <2, [\mu_{\tilde{S}}(2), 1 - \upsilon_{\tilde{S}}(2)]>, \cdots, <n, [\mu_{\tilde{S}}(n), 1 - \upsilon_{\tilde{S}}(n)]>\}$.

Let $\mathrm{Supp}^-(\tilde{S}) = \{i \in N \mid \mu_{\tilde{S}}(i) > 0\}$, $\mathrm{Supp}^+(\tilde{S}) = \{i \in N \mid 1 - \upsilon_{\tilde{S}}(i) > 0\}$, $|\tilde{S}|^-$ and $|\tilde{S}|^+$ represent the number of elements in the support set $\mathrm{Supp}^-(\tilde{S})$ and $\mathrm{Supp}^+(\tilde{S})$ respectively. If $i \in \mathrm{Supp}^-(\tilde{S})$, then $\mu_{\tilde{S}}(i) > 0$ and $\mu_{\tilde{S}}(i) = \mu(i)$; If $i \in \mathrm{Supp}^+(\tilde{S})$, then $1 - \upsilon_{\tilde{S}}(i) > 0$ and $\upsilon_{\tilde{S}}(i) = \upsilon(i)$, where $\mu(i)$ and $\upsilon(i)$ are constant, $\mu(i) \in (0, 1]$, $\upsilon(i) \in (0, 1]$, $\mu(i) + \upsilon(i) \leq 1$. In this paper, the participation and non-participation of member i to intuitionistic fuzzy coalition satisfy the condition that $1 - \upsilon_{\tilde{S}}(i) > 0$ and $\mu_{\tilde{S}}(i) > 0$. Therefore, $\mathrm{Supp}(\tilde{S}) = \mathrm{Supp}^-(\tilde{S}) = \mathrm{Supp}^+(\tilde{S})$, $|\tilde{S}| = |\tilde{S}|^- = |\tilde{S}|^+$. In the cooperation, $\mu(i)$ can be understood as the minimum percentage of the resources invested by the member, $\upsilon(i)$ can be understood as the largest percentage of the resources invested by the member. Therefore, if a member joins a coalitions, its actual participation will be in the range $[\mu(i), 1 - \upsilon(i)]$.

3.3 The Mechanism of Value Creation of Technology Innovation in Construction Industry with Intuitionistic Fuzzy Coalition

The technology innovation coalition of the construction industry converts various elements into innovative technologies by integrating all kinds of human, material and financial resources invested by members. The final innovation results can be new technologies, new processes, new materials and new equipment. The "10 new technologies in construction industry" prepared by Department of Quality and Safety Supervision, Ministry of Housing and Urban Rural Development includes foundation and underground space engineering technology, concrete technology, steel reinforcement and prestress technology, formwork and scaffolding technology, steel structure technology, mechanical and electrical installation technology, green construction technology, waterproof technology, seismic resistance, reinforcement and transformation technology and information application technology. According to different forms of interests, interests can be divided into tangible interests and intangible interests, direct interests and indirect interests, material interests and non-material interests.

Tangible benefit can also be called direct benefit, referring to the benefit brought by the achievements of innovation coalition after solving specific problems, or the benefit brought by the transfer of patented technology created, or the benefit brought by the quantitative production of new materials and new equipment. Tangible benefit are generally material and can be measured in money. Intangible benefit which can also be called indirect benefit generally refers to the improvement of reputation, sustainable development ability the optimization of management experience and business philosophy, as well as the technical know-how and other hidden benefit brought by technological innovation. They can play an active role in the long-term development of the members of the union, which is difficult to be measured with money commonly and belongs to immaterial benefit. In addition to bringing direct and indirect benefit to the members of the coalition, technical innovation in the construction industry can also produce positive social effects. For example, innovation achievements can improve construction quality, reduce pollution, improve the surrounding environment, and promote ecological harmony and social development. This is also the reason why the government is willing to take an active part in the coalition.

4 Shapley Value for Technology Innovation Benefit Distribution in Construction Industry with Intuitionistic Fuzzy Coalition

For the technical innovation coalition of the construction industry, the primary goal is to achieve technological innovation, which can bring a variety of benefit, including direct economic benefit that can be converted into monetary value and indirect benefit that cannot be measured in monetary terms. The technological innovation of the construction industry will have a positive impact on the enterprises, scientific research institutions, relevant universities, government departments and the society involved in the innovation. For the members participating in the coalition, the ultimate goal is to obtain benefit. Therefore, the establishment of a fair and efficient benefit distribution mechanism is the key to the success of the coalition. The essence of benefit distribution problem of construction technology innovation coalition with intuitionistic fuzzy coalition is the solution of intuitionistic fuzzy cooperative game. Shapley value is a solution concept of crisp cooperative games proposed by Shapley in 1953 [37]. It guarantees the fairness and rationality by setting three mathematical axiomatic conditions. It considers the utility distribution result from the perspective of marginal contribution of players. In this paper, the theoretical methods such as intuitionistic fuzzy set, Choquet integral and continuous ordered weighted average operator are used to propose the Shapley value for technology innovation benefit distribution in construction industry with intuitionistic fuzzy coalition.

4.1 The Determination of Benefit of Intuitionistic Fuzzy Coalition Based on Interval Choquet Integral

Definition 2. [37] For any $x \in X$, $\mu_{\bar{A}}$ is said to be a fuzzy set on X, if $\mu_{\bar{A}} : X \to [0,\ 1]$ satisfies $x \in X \mapsto \mu_{\bar{A}}(x) \in [0,\ 1]$. $\bar{A}$ can be shorthand for $\bar{A} = \{<x, \mu_{\bar{A}}(x)> | x \in X\}$. $\mu_{\bar{A}}$ is called to be a the membership function of $\bar{A}$, $\mu_{\bar{A}}(x)$ represents the degree of membership that x belongs to $\bar{A}$. The set of $\bar{A}$ is noted as $G_f(X)$.

Definition 3. [38] Let $\bar{A} \in G_f(X)$ and $\alpha \in [0, 1]$, $\bar{A}_\alpha = \{x \in X \mid \mu_{\bar{A}}(x) \geq \alpha\}$ is called a cut set of $\bar{A}$, α is defined as the confidence level.

In solving practical problems, the concept of cut set is often used to realize the mutual transformation between fuzzy set and classical set.

Definition 4. [38] $\rho : G_O(X) \to [0, +\infty]$ is said to be the fuzzy measure on X, if it satisfies

1. $\rho(\phi) = 0$;
2. $\forall A, B \in G_O(X), A \subseteq B \Rightarrow \rho(A) \leq \rho(B)$.

Definition 5. [39] Let $f : X \to [0, +\infty]$ be a nonnegative bounded measurable function. $\int f d\rho = \int_0^{+\infty} \rho(F_\alpha) d\alpha$ is called to be Choquet integral of f on ρ, where $F_\alpha = \{x \,|\, f(x) \geq \alpha,\ \alpha \in [0, +\infty]\}$..

When $X = (x_1, x_2, \cdots, x_n)$, the corresponding function f can be expressed as $\{f(x_1), f(x_2), \cdots, f(x_n)\}$. Reorder the set of elements $\{x_1, x_2, \cdots, x_n\}$ to $\{x_1^*, x_2^*, \cdots, x_n^*\}$ according to this monotone undiminished sequence; To arrange $f^+(x_i)$ $(i = 1, 2, \cdots, n)$ in an undecreasing monotone order so that $f(x_1^*), f(x_2^*) \leq \cdots \leq f(x_n^*)$. Reorder the set of elements $\{x_1, x_2, \cdots, x_n\}$ to $\{x_1^*, x_2^*, \cdots, x_n^*\}$. At this point, the Choquet integral of the function f can be expressed as:

$$\int f d\rho = \sum_{i=1}^{n} \left[f(x_i^*) - f(x_{i-1}^*)\right] \bullet \rho\left(\{x_i^*, x_{i+1}^*, \cdots, x_n^*\}\right)$$

where $f(x_0^*) = 0$.

In the cooperation game with intuitionistic fuzzy coalition, if a member joins a coalition, the actual participation of the member to the coalition is on the interval $[\mu(i), 1 - \upsilon(i)]$. Let $\bar{u}(i) = [u^-(i), u^+(i)] = [\mu(i), 1 - \upsilon(i)]$, then $\bar{u}(i)$ is an interval-valued fuzzy set. In general, an interval-valued fuzzy set is a bounded closed interval in the set of real numbers, which is commonly expressed by $\bar{x} = [x^-, x^+]$, where x^- and x^+ represent the lower limit and upper limit of an interval-valued fuzzy set respectively. When $x^- = x^+$, $\bar{x}$ becomes a real number, that is, the real number is a special case of the interval-valued fuzzy set. IR denotes the set of all the interval-valued fuzzy sets.

Definition 6. [40] Let $\bar{f} : X \to IR^+$ be a nonnegative bounded measurable function on $X \cdot \int \bar{f} d\rho = \left[\int f^- d\rho, \int f^+ d\rho\right]$ is called to be interval Choquet integral of $\bar{f}$ on X.

According to Definition 5, $\int f^- d\rho = \int_0^{+\infty} \rho(F_\alpha^-)d\alpha \cdot \int f^+ d\rho = \int_0^{+\infty} \rho\left(F_\beta^+\right)d\beta$.

$F_\alpha^- = \{x|f^-(x) \geq \alpha, \alpha \in [0, +\infty]\}$. $F_\beta^+ = \{x|f^+(x) \geq \beta, \beta \in [0, +\infty]\}$.

When $X = (x_1, x_2, \cdots, x_n)$, the corresponding function f^- can be expressed as $f^-(x_1), f^-(x_2), \cdots, f^-(x_n)$. Similarly, f^+ can be expressed as $f^+(x_1), f^+(x_2), \cdots, f^+(x_n)$. To arrange $f^-(x_i)$ $(i = 1, 2, \cdots, n)$ in an undecreasing monotone order so that $f^-(x_1^*) \leq f^-(x_2^*) \leq \cdots \leq f^-(x_n^*)$. Reorder the set of elements $\{x_1, x_2, \cdots, x_n\}$ to $\{x_1^*, x_2^*, \cdots, x_n^*\}$ according to this monotone undiminished sequence; To arrange $f^+(x_i)$ $(i = 1, 2, \cdots, n)$ in an undecreasing monotone order so that $f^+(x_1^{**}) \leq f^+(x_2^{**}) \leq \cdots \leq f^+(x_n^{**})$. Reorder the set of elements $\{x_1, x_2, \cdots, x_n\}$ to $\{x_1^{**}, x_2^{**}, \cdots, x_n^{**}\}$ according to this monotone undiminished sequence $\{x_1^{**}, x_2^{**}, \cdots, x_n^{**}\}$. At this point, the Choquet integral of the function f^- and f^+ can be expressed as follows.

$$\int f^- \mathrm{d}\rho = \sum_{i=1}^{n} [f^-(x_i^*) - f^-(x_{i-1}^*)]\rho(\{x_i^*, x_{i+1}^*, \cdots, x_n^*\})$$

$$\int f^+ \mathrm{d}\rho = \sum_{i=1}^{n} [f^+(x_i^{**}) - f^+(x_{i-1}^{**})]\rho(\{x_i^{**}, x_{i+1}^{**}, \cdots, x_n^{**}\})$$

where $f^-(x_0^*) = 0$ and $f^+(x_0^{**}) = 0$.

Definition 7. For any $\tilde{S} \in G_{if}(N)$, $Q^-(\tilde{S}) = \{\mu_{\tilde{S}}(i)|\mu_{\tilde{S}}(i) > 0, i \in N\}$, let $q^-(\tilde{S})$ represent the number of elements in $Q^-(\tilde{S})$. When $Q^+(\tilde{S}) = \{1 - \upsilon_{\tilde{S}}(i)|1 - \upsilon_{\tilde{S}}(i) > 0, i \in N\}$, let $q^+(\tilde{S})$ represent the number of elements in $Q^+(\tilde{S})$. Arrange the elements in $Q^-(\tilde{S})$ in monotonically increasing order as $h_1^- < h_2^- < \cdots < h_{q^-(\tilde{S})}^-$, Arrange the elements in $Q^+(\tilde{S})$ in monotonically increasing order as $h_1^+ < h_2^+ < \cdots < h_{q^+(\tilde{S})}^+$. For any $\tilde{S} \in G_{if}(N)$, The following equation can be obtained.

$$\begin{aligned}\bar{\gamma}(\tilde{S}) &= [\gamma^-(\tilde{S}), \gamma^+(\tilde{S})] \\ &= \left[\sum_{l=1}^{q^-(\tilde{S})} v\left([\tilde{S}]_{h_l^-}\right)(h_l^- - h_{l-1}^-), \sum_{l=1}^{q^+(\tilde{S})} v\left([\tilde{S}]_{h_l^+}\right)(h_l^+ - h_{l-1}^+)\right]\end{aligned} \tag{1}$$

$\bar{\gamma}$ is called the interval Choquet integral expression of the payment function of cooperative games with intuitionistic fuzzy coalition. In the formula, $h_0^- = 0$, $h_0^+ = 0$. $[\tilde{S}]_{h_l^-} = \{i|\mu_{\tilde{S}}(i) = \mu(i) \geq h_l^-, i \in N\}$ represents the set of all members with minimum participation in coalition $\tilde{S}$ not less than h_l^-, $[\tilde{S}]_{h_l^+} = \{i|\mu_{\tilde{S}}(i) = 1 - \upsilon(i) \geq h_l^+, i \in N\}$ represents the set of all members with minimum participation in coalition $\tilde{S}$ not less than h_l^+. Both $[\tilde{S}]_{h_l^-}$ and $[\tilde{S}]_{h_l^+}$ are crisp coalitions. v represents the payment function of crisp cooperative games.

4.2 The Benefit Determination of Intuitionistic Fuzzy Coalition Based on Continuous Order Weighted Average Operator

In 1988, a famous American scholar *Yager* proposed an ordered weighted average operator in literature, which can carry out weighted average after ordering a group of discrete real Numbers. In 2004, a continuous ordered weighted average operator was further proposed in literature [41], so as to aggregate interval type continuous variable values. The benefit value of n-person cooperative intuitionistic fuzzy coalition based on interval Choquet integral is a continuous interval variable. Therefore, the continuous ordered weighted average operator can be used to aggregate the union benefit with interval values into a single real value.

A function $Q:[0,1]\rightarrow[0,1]$ is called to be basic unit interval monotone function if it satisfies $Q(0)=0$; $Q(1)=1$; for any $x,y\in[0,1]$, if $x>y$, then $Q(x)\geq Q(y)\cdot\Gamma$ is defined as the set of $Q\cdot Q(y)=y^r$, $Q(y)=(\sin(\pi y/2))^r, r>0$ are commonly used.

Definition 8. Let $Q_1,Q_2\in\Gamma\cdot Q_1\geq Q_2$ if $Q_1(y)\geq Q_2(y),\forall y\in[0,1]$.

Definition 9. For $Q\in\Gamma$, $y\in[0,1]$, $F_Q([a,b])=\int_0^1\frac{dQ(y)}{dy}[b-y(b-a)]dy$ is called a continuous ordered weighted average operator.

Theorem 1. If $a_1\geq a_2$, $b_1\geq b_2$, then for any $Q\in\Gamma$, $F_Q([a_1,b_1])\geq F_Q([a_2,b_2])$.

The theorem shows that the continuous ordinal weighted average operator is monotonically non-decreasing with respect to the aggregation interval when the basic unit interval monotone function is invariant. The larger the value of the assembled interval, the larger the real number will be.

Theorem 2. If $Q_1\geq Q_2$, then $F_{Q_1}([a,b])\geq F_{Q_2}([a,b])$.

The theorem shows that the continuous order weighted average operator is monotonically non-decreasing with respect to the monotone function of the basic unit interval when the values of the aggregation interval are invariant. The larger the basic unit interval monotone function is, the larger the real number will be.

Theorem 3. For any $Q\in\Gamma$, $a\leq F_Q([a,b])\leq b$.

The result of the aggregation of continuous variable value $[a,b]$ will not exceed the range of itself.

Definition 10. Let $Q\in\Gamma$ and $y\in[0,1]$. λ is called to be the attitude factor of Q, if $\lambda=\int_0^1 Q(y)dy$.

It can be seen that the attitude factor of each player is determined by the monotone function of its basic unit interval. The attitude factor is the part of the coordinate graph below the range line $Q(y)$ and above the horizontal axis X, where $y\in[0,1]\cdot\lambda\in[0,1]$, which is a real number.

Theorem 4. If $Q\in\Gamma$, λ is the attitude factor of Q, then $F_Q([a,b])=\lambda b+(1-\lambda)a$.

It can be seen that when COWA operator F_Q is used to aggregate the interval $[a,b]$, the result $F_Q([a,b])$ is essentially a weighted average of the left and right endpoints of the interval $[a,b]$, with the weights being $1-\lambda$ and λ.

If the BUM function of member i is Q_i $(i = 1, 2, \cdots, n)$, then its attitude factor is $\lambda_i = \int_0^1 Q_i(y)dy$. The attitude factors of any intuitionistic fuzzy coalition $\tilde{S} \in G_{if}(N)$ are defined as follows.

Definition 11. The attitude factor $\lambda_{\tilde{S}}$ of intuitionistic fuzzy coalition $\tilde{S}$ is defined as the solution of the following optimization problem.

$$\min_{\lambda_{\tilde{S}}}\{\sum_{i \in \text{Supp}(\tilde{S})} (\lambda_{\tilde{S}} - \lambda_i)^2\}$$

Take the derivative of the above expression with respect to λ, and get

$$\lambda_{\tilde{S}} = \sum_{i \in \text{Supp}(\tilde{S})} \lambda_i / |\tilde{S}| \tag{2}$$

Formula (2) shows that the attitude factor of intuitionistic fuzzy coalition is the arithmetic mean of the attitude factor of the members. Therefore, the expression form of interval Choquet integral of intuitionistic fuzzy coalition interests in Definition 4 can be aggregated into real numbers through attitude factor, i.e.

$$\begin{aligned} \gamma(\tilde{S}) &= \lambda_{\tilde{S}}\gamma^+(\tilde{S}) + (1 - \lambda_{\tilde{S}})\gamma^-(\tilde{S}) \\ &= \lambda_{\tilde{S}} \sum_{l=1}^{q^+(\tilde{S})} v\left([\tilde{S}]_{h_l^+}\right)(h_l^+ - h_{l-1}^+) + (1 - \lambda_{\tilde{S}}) \sum_{l=1}^{q^-(\tilde{S})} v\left([\tilde{S}]_{h_l^-}\right)(h_l^- - h_{l-1}^-) \end{aligned} \tag{3}$$

Obviously, if the payment function v of crisp cooperative game satisfies $v(\varnothing) = 0$, then $\gamma(\varnothing) = 0$.

4.3 Shapley Value for Intuitionistic Fuzzy Cooperative Game

Definition 12. For an intuitionistic fuzzy cooperative game (N, γ), let $\tilde{T} \in G_{if}(N)$, $\tilde{T}$ is called a carrier of intuitionistic fuzzy cooperative game (N, γ) if

$$\gamma(\tilde{S} \cap \tilde{T}) = \gamma(\tilde{S}) \; (\tilde{S} \in G_{if}(N))$$

According to the axiomatic method of Shapley value for crisp cooperative game proposed by Shapley [42], and the axiomatic conditions that Shapley value for fuzzy cooperative game with Choquet integral expression of the payment function defined by Tsurumi [43] should meet, the following theorem holds.

Theorem 5. Let $\tilde{S} \in G_{if}(N)$, $\text{Supp}^-(\tilde{S}) = \{i \in N | \mu_{\tilde{S}}(i) > 0\}$, $\text{Supp}^+(\tilde{S}) = \{i \in N | 1 - \upsilon_{\tilde{S}}(i) > 0\}$, $|\tilde{S}|^-$ and $|\tilde{S}|^+$ respectively represents the number of elements in support set $\text{Supp}^-(\tilde{S})$ and $\text{Supp}^+(\tilde{S})$. And it satisfies the condition that if

$1 - \upsilon_{\tilde{S}}(i) > 0$, then, $\mu_{\tilde{S}}(i) > 0$. Therefore, $\mathrm{Supp}(\tilde{S}) = \mathrm{Supp}^-(\tilde{S}) = \mathrm{Supp}^+(\tilde{S})$ and $|\tilde{S}| = |\tilde{S}|^- = |\tilde{S}|^+$. Define a function $\psi_i(\gamma) : G_{if}(N) \to [0, +\infty]$ by

$$\psi_i(\gamma) = \sum_{\substack{i \in \mathrm{Supp}(\tilde{S}) \\ \tilde{S} \in G_{if}(N)}} \frac{(|\tilde{S}| - 1)!(n - |\tilde{S}|)!}{n!} (\gamma(\tilde{S}) - \gamma(\tilde{S} \backslash \{<i, \mu(i), \upsilon(i)>\})) \quad (4)$$

$\psi(\gamma) = (\psi_1(\gamma), \psi_2(\gamma), \cdots, \psi_n(\gamma))$ is called to be Shapley value (or function) for cooperative game (N, γ) with intuitionistic fuzzy coalition.

For the intuitionistic fuzzy cooperative game (N, γ), its Shapley function $\psi_i(\gamma) : G_{if}(N) \to [0, +\infty]$ satisfies the following axioms.

Axiom 1. Effectiveness: If $\tilde{T}$ is a carrier of intuitionistic fuzzy cooperative game (N, γ), then

$$\sum_{i \in \mathrm{Supp}(\tilde{T})} \psi_{if}(\gamma) = \gamma(\tilde{T})$$

Proof. Effectiveness: Let $\tilde{T}$ be a carrier of cooperative game (N, γ) with intuitionistic fuzzy coalition, it can be known from the property (2) of the carrier that $\gamma(\tilde{S} \cup \{<i, \mu(i), \upsilon(i)>\}) = \gamma(\tilde{S})$, where $i \notin \mathrm{Supp}(\tilde{T})$ and $\tilde{S} \in G_{if}(N)$. Then for any $i \notin \mathrm{Supp}(\tilde{T})$, the following formula can be obtained.

$$\begin{aligned}\psi_i(\gamma) &= \sum_{\substack{i \in \mathrm{Supp}(\tilde{S} \cup \{<i, \mu(i), \upsilon(i)>\}) \\ \tilde{S} \in G_{if}(N)}} \frac{[(|\tilde{S}| + 1) - 1]![n - (|\tilde{S}| + 1)]!}{n!} (\gamma(\tilde{S} \cup \{<i, \mu(i), \upsilon(i)>\}) - \gamma(\tilde{S})) \\ &= 0\end{aligned}$$

Therefore

$$\gamma(\tilde{N}) = \sum_{i \in N} \psi_i(\gamma) = \sum_{i \in \mathrm{Supp}(\tilde{T})} \psi_i(\gamma)$$

And we know from Definition 7 that $\gamma(\tilde{T}) = \gamma(\tilde{T} \cap \tilde{N}) = \gamma(\tilde{N})$, then

$$\sum_{i \in \mathrm{Supp}(\tilde{T})} \psi_i(\gamma) = \gamma(\tilde{T})$$

Axiom 2. Symmetry: If π is a permutation of $\tilde{N}$, and $\gamma(\pi\tilde{S}) = \gamma(\tilde{S})$ ($\tilde{S} \in G_{if}(N)$), then

$$\psi_{\pi i}(\gamma) = \psi_i(\gamma)$$

Proof. Symmetry: Let π be a permutation on N which satisfies $\gamma(\pi\tilde{S}) = \gamma(\tilde{S})$, then $|\pi\tilde{S}| = |\tilde{S}|$,

$$\begin{aligned}\psi_{\pi i}(\gamma) &= \sum_{\substack{i \in \mathrm{Supp}(\pi\tilde{S}) \\ \tilde{S} \in G_{if}(N)}} \frac{(|\pi\tilde{S}|-1)!(n-|\pi\tilde{S}|)!}{n!}(\gamma(\pi\tilde{S}) - \gamma(\pi\tilde{S}\backslash\{<\pi i, \mu(i), \upsilon(i)>\})) \\ &= \sum_{\substack{i \in \mathrm{Supp}(\tilde{S}) \\ \tilde{S} \in G_{if}(N)}} \frac{(|\tilde{S}|-1)!(n-|\tilde{S}|)!}{n!}(\gamma(\tilde{S}) - \gamma(\tilde{S}\backslash\{<i, \mu(i), \upsilon(i)>\})) = \psi_i(\gamma)\end{aligned}$$

Axiom 3. Additivity: For (N, γ_1) and (N, γ_2), if there is $(N, \gamma_1 + \gamma_2)$ that can make $\tilde{S} \in G_{if}(N)$ satisfy $(\gamma_1 + \gamma_2)(\tilde{S}) = \gamma_1(\tilde{S}) + \gamma_2(\tilde{S})$, then

$$\psi_{if}(\gamma_1 + \gamma_2) = \psi_{if}(\gamma_1) + \psi_{if}(\gamma_2)$$

$(N, \gamma_1 + \gamma_2)$ is a cooperative game with intuitionistic fuzzy coalition composed by (N, γ_1) and (N, γ_2).

Proof. Additivity: If $(N, \gamma_1 + \gamma_2)$ can make $\tilde{S} \in G_{if}(N)$ satisfy $(\gamma_1 + \gamma_2)(\tilde{S}) = \gamma_1(\tilde{S}) + \gamma_2(\tilde{S})$, then

$$\psi_i(\gamma_1 + \gamma_2) = \sum_{\substack{i \in \mathrm{Supp}(\tilde{S}) \\ \tilde{S} \in G_{if}(N)}} \frac{(|\tilde{S}|-1)!(n-|\tilde{S}|)!}{n!}((\gamma_1 + \gamma_2)(\tilde{S}) - (\gamma_1 + \gamma_2)(\tilde{S}\backslash\{<i, \mu(i), \upsilon(i)>\}))$$

$$= \sum_{\substack{i \in \mathrm{Supp}(\tilde{S}) \\ \tilde{S} \in G_{if}(N)}} \frac{(|\tilde{S}|-1)!(n-|\tilde{S}|)!}{n!}[(\gamma_1(\tilde{S}) + \gamma_2(\tilde{S})) - (\gamma_1(\tilde{S}\backslash\{<i, \mu(i), \upsilon(i)>\}) + \gamma_2(\tilde{S}\backslash\{<i, \mu(i), \upsilon(i)>\}))]$$

$$= \sum_{\substack{i \in \mathrm{Supp}(\tilde{S}) \\ \tilde{S} \in G_{if}(N)}} \frac{(|\tilde{S}|-1)!(n-|\tilde{S}|)!}{n!}[(\gamma_1(\tilde{S}) - \gamma_1(\tilde{S}\backslash\{<i, \mu(i), \upsilon(i)>\})) + (\gamma_2(\tilde{S}) - \gamma_2(\tilde{S}\backslash\{<i, \mu(i), \upsilon(i)>\}))]$$

$$= \psi_i(\gamma_1) + \psi_i(\gamma_2)$$

5 An Illustrative Example

The essence of benefit distribution of technology innovation in construction industry with intuitionistic fuzzy coalition is the solution of intuitionistic fuzzy cooperative game. The following is an example to illustrate the application of it.

There are three members (construction enterprises, scientific research institutions or universities), which are numbered as member 1, 2 and 3. Member i ($i = 1, 2, 3$) has 100 units of resource R_i. Member i gets benefit $v(\{i\})$ by using 100 units of resource R_i to create 100 units of achievement P_i. Any two or three resources between R_1, R_2 and R_3 can create a valuable output. That is, you can use one unit of resource R_i and one unit of resource $R_j(i<j, i,j \in \{1,2,3\})$ to create one unit of outcome P_{ij}. and obtain benefit $v(\{i,j\})$. Moreover, a unit of resource R_1, a unit of resource R_2 and a unit of resource R_3 can jointly create a unit of achievement P_{123} and obtain benefit $v(\{1,2,3\})$.

The interest function v used here is hyper-additive and $v(\varnothing) = 0$. In addition, for $S, T \subseteq \{1,2,3\}$ and $S \cap T = \varnothing$, we have the following relationship $v(S \cup T) \geq v(S) + v(T)$, that is

(1) $P_{123} \geq P_{ij} + P_k$, for $i,j,k \in \{1,2,3\}$ and $i \neq j, j \neq k, k \neq i$;
(2) $P_{ij} \geq P_i + P_j$, for $i,j \in \{1,2,3\}$ and $i \neq j$

As a specific example, the profit value of the coalition is selected as follows (unit: ten thousand Yuan):

$$v(\{1\}) = 100, v(\{2\}) = 200, v(\{3\}) = 200, v(\{1,2\}) = 600$$
$$v(\{2,3\}) = 800, v(\{1,3\}) = 600, v(\{1,2,3\}) = 1200.$$

Under uncertain conditions, the three members plan the resources they could invest in two aspects, that is, the minimum amount of resources to be invested and the amount of resources not to be invested. Member 1 shall invest at least 20 units of resources, and 70 units of resources shall not be invested; Member 2 shall invest at least 40 units of resources, and 40 units of resources shall not be invested; Member 3 shall invest at least 50 units of resources, and 20 units of resources shall not be invested. In addition, the BUM function of member 1 is $Q_1(y) = y$, the BUM function of member 2 is $Q_2(y) = y^2$, and the BUM function of member 3 is $Q_3(y) = y^3$.

$\mu(i)$ represents the ratio of the minimum input resources of a member to the total input resources, $\upsilon(i)$ represents the ratio of the resources that a member must not invest to the total resources. In this case,

$$\mu(1) = 0.2\,, \mu(2) = 0.4\,, \mu(3) = 0.5\,, \upsilon(1) = 0.7\,, \upsilon(2) = 0.4\,, \upsilon(3) = 0.2.$$

Table 1. The interval benefit, attitude factor and aggregation benefit of different intuitionistic fuzzy coalitions

$\tilde{S}$	$\bar{\gamma}(\tilde{S})$	$\lambda_{\tilde{S}}$	$\gamma(\tilde{S})$
$\{<1, 0.2, 0.7>\}$	$[20, 30]$	1/2	25
$\{<2, 0.4, 0.4>\}$	$[80, 120]$	1/3	280/3
$\{<3, 0.5, 0.2>\}$	$[100, 160]$	1/4	115
$\{<1, 0.2, 0.7>, <2, 0.4, 0.4>\}$	$[160, 240]$	5/12	580/3
$\{<1, 0.2, 0.7>, <3, 0.5, 0.2>\}$	$[180, 280]$	3/8	435/2
$\{<2, 0.4, 0.4>, <3, 0.5, 0.2>\}$	$[340, 520]$	7/24	785/2
$\{<1, 0.2, 0.7>, <2, 0.4, 0.4>, <3, 0.5, 0.2>\}$	$[420, 640]$	13/36	4495/9

According to Eqs. (1)–(3), the interval benefit, attitude factors and aggregation benefit of different intuitionistic fuzzy coalitions can be calculated, as shown in Table 1.

According to formula (4), the specific allocation values obtained by each member from different intuitionistic fuzzy coalition can be obtained, as shown in Table 2.

Table 2. Benefit distribution results of different intuitionistic fuzzy coalitions

$\tilde{S}$	Member 1	Member 2	Member 3
$\{<1,0.2,0.7>\}$	25	0	0
$\{<2,0.4,0.4>\}$	0	93.33	0
$\{<3,0.5,0.2>\}$	0	0	115
$\{<1,0.2,0.7>, <2,0.4,0.4>\}$	62.50	130.83	0
$\{<1,0.2,0.7>, <3,0.5,0.2>\}$	63.75	0	153.75
$\{<2,0.4,0.4>, <3,0.5,0.2>\}$	0	185.42	207.08
$\{<1,0.2,0.7>, <2,0.4,0.4>, <3,0.5,0.2>\}$	77.73	199.40	222.31

By analyzing the distribution results in Table 2, it can be seen that Member 1, 2 and 3 will choose the coalition which can make them gain the most, namely the largest coalition, and gain 777,300 Yuan, 199.4 million Yuan and 2,223,100 Yuan respectively.

6 Conclusion

In view of the construction industry is one of the pillar industries of China's national economy, and its development is directly related to the stability. It concerns the economic interests and quality of life of every citizen and should therefore be highly valued. Construction industry technology Innovation strategic alliance is a kind of social organization which serves for the technical innovation of the construction industry. The realization of the construction industry technology innovation, promote the steady development of the construction industry. The realization of the construction industry technology innovation, promote the steady development of the construction industry. In this paper, the concept of intuitionistic fuzzy set is used to describe the forming mechanism and value creation mechanism of the construction industry technology innovation coalition when the members participate in the coalition with a certain degree of hesitation. The essence of this problem is the cooperative game with intuitionistic fuzzy coalition when considering the degree of hesitation of cooperative enterprises. Using mathematical theories and methods such as Choquet integral and continuous ordered weighted average operator, this paper presents Shapley value of cooperative game with intuitionistic fuzzy coalition. It is proved that it can satisfy three axioms, namely, effectiveness, symmetry and additivity, which are similar to the classical Shapley value. Further more, it is applied to the benefit distribution of the construction industry technology innovation with intuitionistic fuzzy coalition.

Intuitionistic fuzzy set is an important generalization of fuzzy set, which has two scale characteristics and can describe the fuzziness more exquisite and effectively, especially the fuzziness of "neither this nor that" type. Therefore, the intuitionistic fuzzy set can be used to describe the fuzzy uncertainty of members in cooperation more effectively. It can reflect the actual problems in a more comprehensive and detailed way, so that participants can make reasonable decisions.

References

1. Wang, Q.X., Zheng, L., Huang, D.A.: Research on innovation in construction industry. Ind. Strateg. **9**, 20–30 (2003). https://doi.org/10.3969/j.issn.0577-7429.2003.09.011
2. Fruin, M., Lin, R., Zhang, H., et al.: Alliance network and innovation: evidence from China's third generation mobile communications industry. J. Asia Bus. Stud. **6**(2), 197–222 (2012). https://doi.org/10.1108/15587891211254407
3. Kangari, R., Miyatake, Y.: Developing and managing innovative construction technologies in Japan. J. Constr. Eng. Manage. **123**(1), 72–78 (1997). https://doi.org/10.1061/(asce)0733-9364(1997)123:1(72)
4. Arditi, D., Kale, S., Tangkar, M.: Innovation in construction equipment and its flow into the construction Industry. J. Constr. Eng. Manage. **123**(4), 371–378 (1997). https://doi.org/10.1061/(asce)0733-9364(1997)123:4(371)
5. Pries, F., Janszen, F.: Innovation in the construction industry: the dominant role of the environment. Constr. Manage. Econ. **13**(1), 43–51 (1995). https://doi.org/10.1080/01446199500000006
6. Eriksson, P.E., Laan, A.: Procurement effects on trust and control in client-contractor relationships. Eng. Constr. Archit. Manage. **14**(4), 387–399 (2016). https://doi.org/10.1016/j.ijpe.2012.01.015
7. Horta, I.M., Camanho, A.S., Costa, J.M.D.: Performance assessment of construction companies: a study of factors promoting financial soundness and innovation in the industry. Int. J. Prod. Econ. **137**(1), 84–93 (2012). https://doi.org/10.1016/j.ijpe.2012.01.015
8. Šuman, N., El-Masr, M.S.: The integrated approach for introducing innovation in construction industry. Organ. Tech. Man. Cons. **5**(2), 834–843 (2013). https://doi.org/10.5592/otmcj.2013.2.2
9. Blayse, A.M., Manley, K.: Key influences on construction innovation. Constr. Innov. **4**(3), 143–154 (2004). https://doi.org/10.1191/1471417504ci073oa
10. Jahn, H., Zimmermann, M., Fischer, M., et al.: Performance evaluation as an influence factor for the determination of profit shares of competence cells in non-hierarchical regional production networks. Robot. CIM-INT Manuf. **22**(5–6), 526–535 (2006). https://doi.org/10.1016/j.rcim.2005.11.011
11. Chauhan, S.S., Proth, J.M.: Analysis of a supply chain partnership with revenue sharing. Int. J. Prod. Econ. **97**(1), 44–51 (2005). https://doi.org/10.1016/j.ijpe.2004.05.006
12. Canakoglu, E., Bilgic, T.: Analysis of a two-stage telecommunication supply chain with technology dependent demand. Eur. J. Oper. Res. **177**(2), 995–1012 (2006). https://doi.org/10.1016/j.ejor.2006.01.006
13. Jia, N.X., Yokoyama, R.: Profit allocation of independent power producers based on cooperative Game theory. Int. J. Elec. Power. **25**(8), 633–641 (2003). https://doi.org/10.1016/s0142-0615(02)00180-1

14. Sakawa, M., Nishizaki, I., Uemura, Y.: Fuzzy programming and profit and cost allocation for a production and transportation problem. Eur. J. Oper. Res. **131**(1), 1–15 (2001). https://doi.org/10.1016/s0377-2217(00)00104-1
15. Zheng, W.J., Zhang, X.M., et al.: Research of profit distributing mechanism of virtual enterprises. J. Ind. Eng. Manage. **15**(1), 26–28 (2001). https://doi.org/10.3969/j.issn.1004-6062.2001.01.008
16. Feng, W.D., Chen, J.: Study on the proportion making for profit/risk allocation within virtual enterprises. Syst. Eng. Theor. Pract. **22**(4), 45–49 (2002). https://doi.org/10.3321/j.issn:1000-6788.2002.04.008
17. Dai, J.H., Xue, H.X.: The strategy of profit allocation among partners in dynamic alliance based on the Shapley value. China J. Manage. Sci. **12**(4), 33–36 (2004). https://doi.org/10.3321/j.issn:1003-207x.2004.04.007
18. Liao, C.L., Fan, Z.J., Tan, A.M.: Research on secondary profit distribution mechanism of virtual enterprises. Sci. Technol. Prog. Policy. **4**, 138–140 (2005). https://doi.org/10.3969/j.issn.1000-7695.2005.04.045
19. Sang, X.M., Li, J.: Profit allocation in small and medium-sized enterprises financing alliance based on interval Shapley Value. Math. Pract. Theory **40**(21), 93–98 (2010)
20. Han, X.L., Zhang, Q.P.: Cooperative game analysis of knowledge creation and benefit distribution among enterprises. Sci. Technol. Prog. Policy **28**(9), 85–87 (2011). https://doi.org/10.3969/j.issn.1001-7348.2011.09.019
21. Chen, M., Yu, J., Zou, H.: Research on profit distribution of supply chain enterprises based on orthogonal projection method. J. Changsha Univ. **25**(5), 99–101 (2011). https://doi.org/10.3969/j.issn.1008-4681.2011.05.034
22. Chen, W., Zhang, Q.: Profit distribution method of enterprise alliance based on fuzzy alliance cooperative game. Trans. Beijing Inst. Technol. **57**(8), 735–739 (2007). https://doi.org/10.3969/j.issn.1001-0645.2007.08.018
23. Tan, C.Q.: Shapley value for n-persons games with interval fuzzy coalition based on Choquet extension. J. Syst. Eng. **25**(4), 451–458 (2010)
24. Atanassov, K.T.: Intuitionistic fuzzy sets. Fuzzy Sets Syst. **20**(1), 87–96 (1986). https://doi.org/10.1016/S0165-0114(86)80034-3
25. Atanassov, K.T.: Intuitionistic Fuzzy Sets. Springer, Heidelberg (1999). https://doi.org/10.1007/978-3-7908-1870-3
26. Li, D.F., Wang, Y.C., Liu, S., et al.: Fractional programming methodology for multi-attribute group decision-making using IFS. Appl. Soft Comput. **9**(1), 219–225 (2009). https://doi.org/10.1016/j.asoc.2008.04.006
27. Li, F.: TOPSIS-based nonlinear-programming methodology for multiattribute decision making with interval-valued intuitionistic fuzzy sets. IEEE Trans. Fuzzy Syst. **18**(2), 299–311 (2010). https://doi.org/10.1109/TFUZZ.2010.2041009
28. Li, D.F., Shan, F., Cheng, C.T.: On properties of four IFS operators. Fuzzy Sets Syst. **154**(1), 151–155 (2005). https://doi.org/10.1016/j.fss.2005.03.004
29. Xu, Y.J., Sun, T., Li, D.F.: Intuitionistic fuzzy prioritized OWA operator and its application in multi-criteria decision-making problem. Control Decis. **26**(1), 129–132 (2011). https://doi.org/10.1111/j.1759-6831.2010.00112.x
30. Xu, X.Y., Huang, X.Y., Zhao, Y., Huang, Y.: Fault diagnosis with empirical model decomposition and intuitionistic fuzzy reasoning neural networks. J. Chongqing Univ. Technol. (Nat. Sci. Ed.) **24**(4), 91–96 (2010). https://doi.org/10.3969/j.issn.1674-8425-b.2010.04.018
31. Li, D.F., Nan, J.X.: A nonlinear programming approach to matrix games with payoffs of Atanassov's intuitionsistic fuzzy sets. Int. J. Uncertain. Fuzziness **17**(04), 585–607 (2009). https://doi.org/10.1142/S0218488509006157

32. Li, D.F.: Mathematical-programming approach to matrix games with payoffs represented by Atanassov's interval-valued intuitionistic fuzzy sets. IEEE Trans. Fuzzy Syst. **18**(6), 1112–1128 (2011). https://doi.org/10.1109/TFUZZ.2010.2065812
33. Nan, J.X., Li, D.F., Zhang, M.J.: A lexicographic method for matrix Games with payoffs of triangular intuitionistic fuzzy numbers. Int. J. Comput. Int. Syst. **3**(3), 280–289 (2010). https://doi.org/10.1080/18756891.2010.9727699
34. Nayak, P.K., Pal, M.: Bi-matrix games with intuitionistic fuzzy goals. Iran. J. Fuzzy Syst. **7** (1), 65–79 (2010). https://doi.org/10.3934/ipi.2010.4.191
35. Angelov, P.P.: Optimization in an intuitionistic fuzzy environment. Fuzzy Sets Syst. **86**(3), 299–306 (1997). https://doi.org/10.1016/s0165-0114(96)00009-7
36. Li, D.F.: Intuitionistic Fuzzy Set Decision and Game Analysis Methodologies. National Defense Industry Press, Beijing (2012)
37. Han, L.Y., Wang, P.Z.: Applied Fuzzy Mathematics. Capital University of Economics and Business Press, Beijing (1998)
38. Sugeno, M.: Theory of fuzzy integral and its applications. Doctor, Tokyo Institute of Technology (1974)
39. Murofushi, T., Sugeno, M.: An interpretation of fuzzy measure and the Choquet integral as an integral with respect to a fuzzy measure. Fuzzy Sets Syst. **29**(1), 201–227 (1989). https://doi.org/10.1016/0165-0114(89)90194-2
40. Yang, R., Wang, Z., Heng, P.A., et al.: Fuzzy numbers and fuzzification of the Choquet integral. Fuzzy Sets Syst. **153**(1), 95–113 (2005). https://doi.org/10.1016/j.fss.2004.12.009
41. Yager, R.R.: OWA aggregation over a continuous interval argument with application to decision making. IEEE Trans. Syst. Man Cybern. Part B (Cybern.) **34**(5), 1952–1963 (2004). https://doi.org/10.1109/tsmcb.2004.831154
42. Shapley, L.S.: A value for n-person games. In: Kuhn, H.W., Tucker, A.W. (eds.) Contributions to the Theory of Games II. Annals of Mathematics Studies, pp. 307–317. Princeton University Press, Princeton (1953)
43. Tsurumi, M., Tanino, T., Inuiguchi, M.: A Shapley function on a class of cooperative fuzzy games. Eur. J. Oper. Res. **129**(3), 596–618 (2001). https://doi.org/10.1016/S0377-2217(99)00471-3

The Existence of the pseu-Ky Fan' Points and the Applications in Multiobjective Games

Xiaoling Qiu(✉)

College of Mathematics and Statistics, Guizhou University, Guiyang 550025, Guizhou, China
xlqiu@gzu.edu.cn

Abstract. In this paper, we first propose pseu-Ky Fan' points for vector Ky Fan inequalities and prove the existence results under some relaxed assumptions by virtue of KKMF principle and Fan-Browder fixed point theorem. Mild continuity named pseudocontinuity is introduced for the existence results which is weaker than semicontinuity and generalizes the present results in the literature. As applications, we define pseu-weakly Pareto-Nash equillibrium for multiobjective games and obtain some existence theorems.

Keywords: vector Ky Fan inequalities · pseu-Ky Fan' points · Pseudocontinuity · KKMF lemma · pseu-weakly Pareto-Nash equillibrium

1 Introduction

Let X be a nonempty set of Hausdorff topological space E and $\Phi : X \times X \to R^n$ be a vector-valued function. The vector Ky Fan inequality which we will deal with is to find $x^* \in X$ such that

$$\Phi(x^*, y) \notin intR^n_+, \qquad \forall y \in X.$$

We call x^* a solution or a Ky Fan point of the vector Ky Fan inequality Φ.

Vector Ky Fan inequalities are natural generalizations of the Ky Fan inequality to vector-valued functions. The important inequality was introduced by Ky Fan [1] which now has been called Ky Fan inequality and the solution x^* of Ky Fan inequality was called Ky Fan's points first by Tan, Yu and Yuan in 1995, to see [2]. It is well known that Ky Fan inequality plays a very important role in many fields such as game theory, fixed point theory, variational inequalities, control theory and mathematical economics, etc. A great deal of fruitful results have been achieved on how to improve and apply the important inequalities such as [3–6] and references therein. [6] proved some existence results of the solutions

Supported by the National Natural Science Foundation of Guizhou Province (Grant No. QKH[2016]7425), and the Scientific Research Projects for the Introduced Talents of Guizhou University (Grant No. [2018]11).

D. Li (Ed.): EAGT 2019, CCIS 1082, pp. 105–115, 2019.
https://doi.org/10.1007/978-981-15-0657-4_7

and the compactness of the solution sets for vector-valued functions with the cone semicontinuity and the cone quasiconvexity in infinite dimensional spaces. [7,8] study the set-valued versions of Ky Fan inequality and [7]established the notion of weakly Ky Fan's points of set-valued mapping and prove some existence theorems of weakly Ky Fan's points while [8] obtained the two set-valued versions of Ky Fan inequalities and deduced Schauder's and Kakutani's fixed point theorems. In this paper, we will show the new version of vector Ky Fan inequalities with pseudocontinuous functions and established some the existence results of pseu-Ky Fan's points, which generalize the present results in [6,7,9]. As applications, we will give the existence results of pseu-weakly Pareto-Nash equilibrium for multiobjective games.

The rest of the paper is organized as follows. In Sect. 2, we first recall some definitions including pseudocontinuity and their properties which are needed in the sequel. In Sect. 3, we propose the concept of pseu-Ky Fan'points and give some existence results for the vector Ky Fan inequalities, then show some other cases of vector Ky Fan inequalities with pseudocontinuity. Finally, we will obtain some existence results of pseu-weakly Pareto-Nash equilibrium for multiobjective games as applications in Sect. 4.

2 Preliminaries

Now let us begin with some definitions and lemmas which we will use in the later.

Definition 2.1 ([10]). Let X be a Hausdorff topological space and $f : X \to R$ be a function.

(1) f is said to be upper pseudocontinuous at $x_0 \in X$ if for all $x \in X$ such that $f(x_0) < f(x)$, we have
$$\limsup_{y \to x_0} f(y) < f(x);$$
(2) f is said to be upper pseudocontinuous on X if it is upper pseudocontinuous at each x of X;
(3) f is said to be lower pseudocontinuous at $x_0 \in X$ if for all $x \in X$ such that $f(x) < f(x_0)$, we have
$$f(x) < \liminf_{y \to x_0} f(y);$$
(4) f is said to be lower pseudocontinuous on X if it is lower pseudocontinuous at each x of X;
(5) f is said to be pseudocontinuous at $x \in X$ if f is both upper pseudocontinuous and lower pseudocontinuous at x; f is said to be pseudocontinuous on X if f is pseudocontinuous at each x of X.

Remark 2.1. If f is upper pseudocontinuous on X, then $-f$ is lower pseudocontinuous on X. The converse is also true.

Remark 2.2. Each upper (resp. lower) semicontinuous function is also upper (resp. lower) pseudocontinuous. But the converse is not true. For example: let $X = [0, 2]$, $f_i : X \to R, i = 1, 2$ be defined as fellows:

$$f_1(x) := \begin{cases} -x, & 0 \le x < 1, \\ -2, & 1 \le x < 2. \end{cases}; \quad f_2(x) := \begin{cases} x, & 0 \le x < 1, \\ 2, & 1 \le x < 2. \end{cases}$$

One can easily check that f_1 is upper pseudocontinuous but not upper semicontinuous at $x = 1$ and that f_2 is not lower semicontinuous but lower pseudocontinuous at $x = 1$.

Lemma 2.1 ([10]). Let X be a Hausdorff topological space and $f : X \to R$ be lower pseudocontinuous, then $\forall b \in f(X)$, the set $\{x \in X : f(x) \le b\}$ is closed.

Definition 2.2. Let X be a nonempty subset of Hausdorff topological space E, and $F = (f_1, \cdots, f_k) : X \to R^k$ be a vector-valued function.

(1) F is said to be lower pseudocontinuous at $x \in X$ if and only if f_i is lower pseudocontinuous at $x \in X$ for any $i = 1, \cdots, k$;
(2) F is said to be lower pseudocontinuous on X if F is lower pseudocontinuous at each x of X;
(3) F is said to be upper pseudocontinuous on X if $-F$ is lower pseudoncontinuous on X.

Lemma 2.2. Let X be a nonempty subset of Hausdorff topological space E. Let the vector function $F : X \to R^k$ be lower pseudocontinuous on X and $F(0) = 0$. Then $G = \{x \in X : F(x) \notin intR^k_+\}$ is a closed set in X.

Proof. Let $F(x) = (f_1(x), \cdots, f_k(x))$, obviously, $G = \bigcup_{i=1}^{k} G_i(x)$, where $G_i(x) = \{x \in X : f_i(x) \le 0\}, \forall i = 1, \cdots, k$. From the Lemma 2.1 and $f_i(x) = 0$, we can see that $G_i(x)$ is a closed set in X. Thus, G is closed in X.

Corollary 2.1. Let X be a nonempty subset of Hausdorff topological space E and the vector function $F : X \to R^k$ be lower pseudocontinuous on X and $F(0) = 0$. Then $G = \{x \in X : F(x) \in intR^k_+\}$ is a open set in X.

Definition 2.3 ([11]). Let X be a nonempty convex subset of Hausdorff topological space E, $f : X \to R$ be the real number function. $\forall x_1, x_2 \in X, \forall \lambda \in (0, 1)$,

(1) f is said to be convex function on X, if there holds

$$f(\lambda x_1 + (1 - \lambda)x_2) \le \lambda f(x_1) + (1 - \lambda)f(x_2);$$

(2) f is said to be concave function on X if $-f$ is convex function on X. That means

$$f(\lambda x_1 + (1 - \lambda)x_2) \ge \lambda f(x_1) + (1 - \lambda)f(x_2);$$

(3) f is said to be quasi-concave function on X if there holds

$$f(\lambda x_1 + (1 - \lambda)x_2) \ge \min\{f(x_1), f(x_2)\};$$

(4) f is said to be quasi-convex function on X if $-f$ is quasi-concave function on X. That means

$$f(\lambda x_1 + (1-\lambda)x_2) \leq \max\{f(x_1), f(x_2)\}.$$

Lemma 2.3. Let X be a nonempty convex subset of Hausdorff topological space E, the real number function $f : X \to R$,

(1) f is quasi-concave function on X if and only if $\forall r \in R, \{x \in X : f(x) > r\}$ is convex;

(2) f is quasi-convex function on X if and only if $\forall r \in R, \{x \in X : f(x) < r\}$ is convex.

Definition 2.4. Let X be a nonempty convex subset of Hausdorff topological space E and $F = (f_1, \cdots, f_k) : X \to R^k$ be a vector-valued function. Then F is said to be R^k_+- quasi-convex on X if and only if f_i is quasi-convex on X for any $i = 1, \cdots, k$ and F is said to be R^k_+-quasi-concave on X if $-F$ is R^k_+-quasi-convex on X.

Lemma 2.4 ([6]). Let X be a convex subset of Hausdorff topological space E, let the vector function $F : X \to R^k$ be R^k_+-quasi-concave on X. Then $G = \{x \in X : F(x) \in intR^k_+\}$ is convex on X.

The following well-known KKMF lemma is an important generalization of KKM theorem to the infinite dimensional space by Ky Fan [12].

Lemma 2.5 (KKMF Lemma). Let X be a nonempty convex subset of Hausdorff topological vector space E, and $F : X \to X$ be a set-valued mapping. For each $x \in X, F(x)$ is closed, and there exists some $x_0 \in X$ such that $F(x_0)$ is compact. If $Co\{x_1, x_2, \cdots, x_n\} \subset \bigcup_{i=1}^n F(x_i)$, where $Co\{x_1, x_2, \cdots, x_n\}$ is the convex hull of $\{x_1, x_2, \cdots, x_n\}$, then $\bigcap_{x \in X} F(x) \neq \emptyset$.

The following fixed theorem is Fan-Browder fixed point theorem.

Lemma 2.6 (Fan-Browder Fixed Point Theorem) ([13]). Let X be a nonempty convex and compact subset of Hausdorff topological vector space E. Suppose a set-valued mapping $F : X \to X$ has the following properties:
(1) $\forall x \in X, F(x)$ is nonempty and convex;
(2) $\forall y \in X$, The inverse valued $F^{-1}(y) = \{x \in X : y \in F(x)\}$ is open in X.
Then F has at least one fixed point.

3 Pseu-Ky Fan's Points

In the previous section, we discussed the existence of the solutions for vector Ky Fan inequalities with pseudocontinuity defined on a compact set.

Theorem 3.1. Let X be a nonempty convex and compact subset of Hausdorff topological space E. The vector function $\Phi(x,y) = (\Phi_1(x,y),\cdots,\Phi_k(x,y)) : X \times X \to R^k$, where $\Phi_j : X \times X \to R$ for any $j = 1,\cdots,k$, is satisfying:

(1) $\forall y \in X, \forall j = 1,\cdots,k, x \to \Phi_j(x,y)$ is lower pseudocontinuous on X;
(2) $\forall x \in X, \forall j = 1,\cdots,k, y \to \Phi_j(x,y)$ is quasi-concave on X;
(3) $\forall x \in X, \forall j = 1,\cdots,k, \Phi_j(x,x) = 0$.

Then there exists $x^* \in X$ such that $\Phi(x^*,y) \notin intR^k_+$ for any $y \in X$.

Proof. For any $y \in X$, denote by $F(y) = \{x \in X : \Phi(x,y) \notin intR^k_+\}$. By (3), $\Phi(y,y) = 0 \notin intR^k_+$, then $y \in F(y)$ and $F(y) \neq \emptyset$. From the above Lemma 2.2, $F(y)$ is compact.

Next we will proof that for any $\{y_1, y_2,\cdots,y_n\} \subset X$, there holds

$$Co\{y_1,y_2,\cdots,y_n\} \subset \bigcup_{i=1}^{n} F(y_i).$$

By the way of contradiction, there exists $y_0 \in Co\{y_1,y_2,\cdots,y_n\} \subset X$ and $y_0 = \sum_{i=1}^{n} \alpha_i y_i$ with $\alpha_i \geq 0, i = 1,2,\cdots,n, \sum_{i=1}^{n} \alpha_i = 1$ but $y_0 \notin \bigcup_{i=1}^{n} F(y_i)$. Then for $i = 1,\cdots,n, y_0 \notin F(y_i)$, i.e., $\Phi(y_0,y_i) \in intR^k_+$.
Hence, for any $j = 1,\cdots,k$, we have $\Phi_j(y_0,y_i) > 0$. By (2), for any $j = 1,\cdots,k$,

$$\Phi_j(y_0,y_0) \geq \min_{1\leq i\leq n} \Phi_j(y_0,y_i) > 0,$$

Which is a contradiction with the condition (3).

Therefore, by KKMF lemma, we know $\bigcap_{y\in X} F(y) \neq \emptyset$. We take $x^* \in \bigcap_{y\in X} f(y)$, which implies

$$\Phi(x^*,y) \notin intR^k_+ \quad \forall y \in X.$$

Remark 3.1. We call x^* as a pseu-Ky Fan's point if x^* is a solution of the vector Ky Fan inequality and the vector function $\Phi(x,y)$ satisfies the condition(1) of Theorem 3.1.

Remark 3.2. We also obtain the existence result by the Fan-Browder fixed point theorem. Assume by contradiction that for any $x \in X$, $F(x) = \{y \in X, \Phi(x,y) \in intR^k_+\} \neq \emptyset$. By (2), we know $F(x)$ is convex. For any $y \in X, F^{-1}(y) = \{x \in X, \Phi(x,y) \in intR^k_+\}$. By Corollary 2.1, $F^{-1}(y)$ is open. These are sufficient condition for Fan-Browder fixed point theorem. Thus, there exists $x^* \in X$ such that $x^* \in F(x^*)$. That means $\Phi(x^*,x^*) \in intR^k_+$, which is in contradiction with the condition (3).

Remark 3.3. If $k = 1$, the vector Ky Fan inequality reduce the Ky Fan inequality as follows, to see [9].

Let X be a nonempty convex and compact subset of Hausdorff topological space E. The function $\phi(x,y) = X \times X \to R$, is satisfying:
(1) $\forall y \in X, x \to \phi(x,y)$ is lower pseudocontinuous on X;
(2) $\forall x \in X, y \to \phi(x,y)$ is quasi-concave on X;
(3) $\forall x \in X, \phi(x,x) = 0$.
Then there exists $x^* \in X$ such that $\phi(x^*, y) \leq 0$ for any $y \in X$.

That means x^* is a pseu-Ky Fan's point of the function $\phi(x,y)$.

Let X be a nonempty set of Hausdorff topological space E, and $\Psi : X \times X \to R^n$ be a vector-valued function. If there exists $x^* \in X$ such that

$$\Psi(x^*, y) \notin -intR^k_+, \qquad \forall y \in X.$$

Then x^* is called a solution of the vector equilibrium problem.

From the above Theorem 3.1, we can obtain an existence result of the solution for vector equilibrium problem as follows.

Theorem 3.2. Let X be a nonempty convex and compact subset of Hausdorff topological space E. The vector function $\Psi(x,y) = (\Psi_1(x,y), \cdots, \Psi_k(x,y)) : X \times X \to R^k$, where $\Psi_i : X \times X \to R, \forall i = 1, \cdots, k$, is satisfying:

(1) $\forall y \in X, \forall j = 1, \cdots, k, x \to \Psi_j(x,y)$ is upper pseudocontinuous on X;
(2) $\forall x \in X, \forall j = 1, \cdots, k, y \to \Psi_j(x,y)$ is quasi-convex on X;
(3) $\forall x \in X, \forall j = 1, \cdots, k, \Psi_j(x,x) = 0$.

Then there exists $x^* \in X$ such that $\Psi(x^*, y) \notin -intR^k_+$ for any $y \in X$.

Proof. $\forall x \in X, \forall y \in X$, Set $\Phi(x,y) = -\Psi(x,y)$. It is easy to check that

(1) $\forall y \in X, \forall j = 1, \cdots, k, x \to \Phi_j(x,y)$ is lower pseudocontinuous on X;
(2) $\forall x \in X, \forall j = 1, \cdots, k, y \to \Phi_j(x,y)$ is quasi-concave on X;
(3) $\forall x \in X, \forall j = 1, \cdots, k, \Phi_j(x,x) = 0$.

By Theorem 3.1, there exists $x^* \in X$ such that $\Phi(x^*, y) \notin intR^k_+$ for any $y \in X$. That implies $\Psi(x^*, y) \notin -intR^k_+$ for any $y \in X$.

Remark 3.4. Theorem 3.1 is shown the new expression on the vector Ky Fan inequality without the R^k_+-lower semicontinuity and R^k_+ quasi-concavity of the vector function Φ on the cone. The quasi-concavity is just needed for the every component of the vector function Φ and the lower pseudocontinuity which is weaker than low semicontinuity is required for the components. So our judgment will be direct from the component of the vector function Φ.

Remark 3.5. Similarly, Theorem 3.2 is shown the new expression on the vector equilibrium inequality without the R^k_+-upper semicontinuity and R^k_+-quasi-convexity of the vector function Ψ on the cone. The quasi-concavity and the lower pseudocontinuity are just needed for the component of the vector function Ψ for the existence of the solution.

In the above theorems, we discussed the existence of the pseu-Ky Fan's points of the vector Ky Fan inequalities in the case of nonempty convex compact set.

Theorem 3.3. Let X be a nonempty unbounded closed convex subset of Hausdorff linear topological space E, $\phi : X \times X \to R^k$ be a vector function with $\Phi(x,y) = (\Phi_1(x,y), \cdots, \Phi_k(x,y))$ where $\Phi_i : X \times X \to R$ for any $i = 1, \cdots, k$. Suppose Φ satisfies the following conditions:

(1) $\forall y \in X, \forall i = 1, \cdots, k, x \to \Phi_i(x,y)$ is pseudocontinuous on X;
(2) $\forall x \in X, \forall i = 1, \cdots, k, y \to \Phi_i(x,y)$ is quasi-concave on X;
(3) $\forall x \in X, \Phi_i(x,x) = 0$;
(4) for any sequence $\{x^m\}$ with $\|x^m\| \to \infty$, there exist a positive integer m_0 and $y \in X$ such that $\|y\| \leq \|x^m\|$ and $\Phi(x^{m_0}, y) \in intR^k_+$.

Then there exists $x^* \in X$ such that $\Phi(x^*, y) \notin intR^k_+$ for any $y \in X$.

Proof. For each $m = 1, 2, \cdots$, set $C_m = \{x \in X : \|x\| \leq m\}$. We may assume that $C_m \neq \emptyset$. C_m is a bounded closed convex subset in X since X is closed convex set. By Theorem 3.1, there exist $x^m \in C_m$ such that $\Phi(x^m, y) \notin intR^k_+$ for any $y \in C_m$.

If the sequence $\{x^m\}$ is unbounded in X, we can suppose that $\|x^m\| \to \infty$ (otherwise subsequence). By (4), there exist a positive integer m_0 and $y \in X$ such that $\|y\| \leq \|x^{m_0}\|$ and $\Phi(x^{m_0}, y) \in intR^k_+$, which is a contradiction with $\|y\| \leq \|x^{m_0}\| \leq m_0, y \in C_{m_0}, \Phi(x^{m_0}, y) \notin intR^k_+$. Thus $\{x^m\}$ is bounded in X and there is M such that $\|x^m\| \leq M$. Since C_M is bounded and we may assume $x^m \to x^* \in C_M \subset X$.

$\forall y \in X$, There is a positive integer K such that $y \in C_k$ and $C_k \subset C_M, y \in C_M, \Phi(x^m, y) \notin intR^k_+$ when $m \geq k$. Set $F(y) = \{x \in X : \Phi(x,y) \notin intR^k_+\}$, by (1) and Corollary 2.1, then $F(y)$ is closed. Since $x^m \to x^*$, then $x^* \in F(y)$ which implies $\Phi(x^*, y) \notin intR^k_+$. The proof is thus complete.

Remark 3.6. Theorem 3.3 is shown the existences of the solution of the new version on vector Ky Fan inequalities with pseudocontinuity in the unbounded setting [14].

4 Applications

Let $N = \{1, 2, \cdots, n\}$ be the set of players. For each $i \in N$, let X_i be the strategy set for player i, $X = \prod_{i=1}^n X_i$, $F^i = (f_1^i, \cdots, f_k^i) : X \to R^k$ be the vector-valued payoff of player i, where k is a positive integer. This multiobjective game is denoted by $\Gamma^* = \{X_i, F^i\}_{i \in N}$. A strategy profile $x^* = (x_1^*, x_2^*, \cdots, x_n^*) \in X$ is called a weakly Pareto-Nash equilibrium of a multiobjective game Γ^* if for each $i \in N$,

$$F^i(y_i, x_{\hat{i}}^*) - F^i(x_i^*, x_{\hat{i}}^*) \notin \text{int}R^k_+, \ \forall y_i \in X_i,$$

where $x_{\hat{i}}^* = (x_1^*, \cdots, x_{i-1}^*, x_{i+1}^*, \cdots, x_n^*)$.

For any $i \in N$, if F^i has some pseucontinuous property, we say a strategy x^* is a pseu-weakly Pareto-Nash equilibrium for multiobjective games.

From the above theorems, we can obtain some existence results of a pseu-weakly Pareto-Nash equilibrium of multiobjective games as applications.

Theorem 4.1. Let the multiobjective game $\Gamma^* = \{X_i, F^i\}_{i\in N}$ satisfy the following conditions:

(1) $\forall i \in N$, X_i is a nonempty, convex and compact subset of a Hausdorff topological space E_i;

(2) $\forall j = 1, 2, \cdots, k$, $\forall y = (y_1, \cdots, y_n) \in X, x \to \sum_{i=1}^{n}[f_j^i(y_i, x_{\hat{i}}) - f_j^i(x_i, x_{\hat{i}})]$ is lower pseudocontinuous on X;

(3) $\forall j = 1, 2, \cdots, k$, $\forall x = (x_1, \cdots, x_n) \in X, y \to \sum_{i=1}^{n}[f_j^i(y_i, x_{\hat{i}}) - f_j^i(x_i, x_{\hat{i}})]$ is quasi-concave on X;

Then there exists at least a pseu-weakly Pareto-Nash equilibrium of Γ^*.

Proof. $\forall x = (x_1, \cdots, x_n) \in X, \forall y = (y_1, \cdots, y_n) \in X$, Denote by

$$\Phi(x, y) = \sum_{i=1}^{n}[F_i(y_i, x_{\hat{i}}) - F_i(x_i, x_{\hat{i}})].$$

We notice that $\forall j = 1, \cdots, k, \Phi_j(y, y) = 0$ for any $y \in X$. By the conditions (1)(2) and Theorem 3.1, there exists $x^* \in X$ such that $\Phi(x^*, y) \notin intR_+^k$ for any $y \in X$. $\forall i \in N, \forall y_i \in X_i$, Set $y = (y_i, x_{\hat{i}}^*)$, then $y \in X$.

Then

$$\Phi(x^*, y) = \sum_{i=1}^{n}[F_i(y_i, x_{\hat{i}}) - F_i(x_i, x_{\hat{i}})] = F_i(y_i, x_{\hat{i}}^*) - F_i(x_i^*, x_{\hat{i}}^*) \notin intR_+^k,$$

which implies $x^* \in X$ is a pseu-weakly Pareto-Nash equilibrium of a multiobjective game Γ^*.

Assume that $k_1 \le k_2 \le \cdots \le k_n$, by applying Theorem 4.1, we obtain the following existence theorem.

Theorem 4.2. Let $\Gamma^{**} = \{X_i, F^i\}_{i\in N}$ be the multiobjective game, where $F^i = (f_1^i, \cdots, f_{k_i}^i)$. Suppose that Γ^{**} satisfies the following conditions:

(1) $\forall i \in N$, X_i is a nonempty, convex and compact subset of a Hausdorff topological space E_i;

(2) $\forall j = 1, 2, \cdots, k_i$, $\forall y = (y_1, \cdots, y_n) \in X, x \to \sum_{i=1}^{n}[f_j^i(y_i, x_{\hat{i}}) - f_j^i(x_i, x_{\hat{i}})]$ is lower pseudocontinuous on X;

(3) $\forall j = 1, 2, \cdots, k_i$, $\forall x = (x_1, \cdots, x_n) \in X, y \to \sum_{i=1}^{n}[f_j^i(y_i, x_{\hat{i}}) - f_j^i(x_i, x_{\hat{i}})]$ is quasi-concave on X;

Then there exists a pseu-weakly Pareto-Nash equilibrium of Γ^{**}.

Proof. $\forall x = (x_1, \cdots, x_n) \in X, \forall y = (y_1, \cdots, y_n) \in X$, We define the vector-valued function $\Psi : X \times X \to R^{k_n}$ by

$$\Psi(x,y) = \sum_{i=1}^{n} \Psi_i(x,y)$$

where

$$\Psi_i(x,y) = (\underbrace{F^i(y_i, x_{\hat{i}}) - F^i(x_i, x_{\hat{i}})}_{k_i \text{ components}}, \underbrace{\sum_{i=1}^{n}[f_1^i(y_i, x_{\hat{i}}) - f_1^i(x_i, x_{\hat{i}})], \cdots, \sum_{i=1}^{n}[f_1^i(y_i, x_{\hat{i}}) - f_1^i(x_i, x_{\hat{i}})])}_{k_n - k_i \text{ components}}$$

It is easy to check that

(1) $\forall y = (y_1, \cdots, y_n) \in X, x \to \Psi(x,y)$ is lower pseudocontinuous on X;
(2) $\forall x = (x_1, \cdots, x_n) \in X, y \to \Psi(x,y)$ is quasi-concave on X;
(3) $\forall x \in X, \Psi(x,x) = 0 \notin intR_+^{k_n}$.

Therefore, by Theorem 3.3, there exists $x^* \in X$ such that $\Psi(x^*, y) \notin intR_+^{k_n}$ for any $y \in X$. For each $i \in N$ and $y_i \in X_i$, set $y = (y_i, x_{\hat{i}}^*) \in X$, then

$$\Psi_i(x^*, y) = \Psi(x^*, y) \notin intR_+^{k_n}.$$

If $F^i(y_i, x_{\hat{i}}^*) - F^i(x_i^*, x_{\hat{i}}^*) \in intR_+^{k_i}$, then $f_j^i(y_i, x_{\hat{i}}^*) - f_j^i(x_i^*, x_{\hat{i}}^*) \in intR_+$, for each $j = 1, \cdots, k_i$ and $\Psi_i(x^*, y) \in intR_+^{k_n}$, which contradicts that $\Psi_i(x^*, y) \notin intR_+^{k_n}$. Hence, $F^i(y_i, x_{\hat{i}}^*) - F^i(x_i^*, x_{\hat{i}}^*) \notin intR_+^{k_i}$, for each $i \in N$, i.e., x^* is a pseu-weakly Pareto-Nash equilibrium point of the multiobjective game Γ^{**}. The proof is completed.

Theorem 4.3. $\forall i \in N$, Let X_i is a nonempty closed convex subset of Hausdorff linear topological space E_i, $X = \Pi_{i=1}^n$, $F^i = \{f_1^i, \cdots, f_k^i\} : X \to R^k$ satisfy the following conditions:

(1) $\forall j = 1, 2, \cdots, k$, $\forall y = (y_1, \cdots, y_n) \in X, x \to \sum_{i=1}^{n}[f_j^i(y_i, x_{\hat{i}}) - f_j^i(x_i, x_{\hat{i}})]$ is lower pseudocontinuous on X;
(2) $\forall j = 1, 2, \cdots, k$, $\forall x = (x_1, \cdots, x_n) \in X, y \to \sum_{i=1}^{n}[f_j^i(y_i, x_{\hat{i}}) - f_j^i(x_i, x_{\hat{i}})]$ is quasi-concave on X;
(3) $\forall x \in X, \phi_i(x,x) = 0$;
(4) For any sequence $\{x^m = (x_1^m, \cdots, x_n^m)\}$ with $\|x^m\| = \sum_{i=1}^{n} \|x_i^m\|_i \to \infty$, where $\|x_i^m\|_i$ means the norm of x_i^m in X_i, there exist some $i \in N$, a positive integer m_0 and $y \in X$ such that $\|y_i\| \leq \|x_i^m\|_i$ and $F^i(y_i, x_{\hat{i}}^{m_0}) - F^i(x_i^{m_0}, x_{\hat{i}}^{m_0}) \in intR_+^k$.

Then there exists a pseu-weakly Pareto-Nash equilibrium of multiobjective game.

Proof. $\forall x = (x_1, \cdots, x_n) \in X, \forall y = (y_1, \cdots, y_n) \in X$, Denote by

$$\phi(x,y) = \sum_{i=1}^{n} [F^i(y_i, x_{\hat{i}}) - F^i(x_i, x_{\hat{i}})].$$

It is easy to check

(1) $\forall y \in X, \forall i = 1, \cdots, k, x \to \phi_i(x,y)$ is pseudocontinuous on X;
(2) $\forall x \in X, \forall i = 1, \cdots, k, y \to \phi_i(x,y)$ is quasi-concave on X;
(3) $\forall x \in X, \phi_i(x,x) = 0$;

By (4) , for any sequence $\{x^m = (x_1^m, \cdots, x_n^m)\}$ with $\|x^m\| = \sum_{i=1}^{n} \|x_i^m\|_i \to \infty$, there exist some $i \in N$, a positive integer m_0 and $y \in X$ such that $\|y_i\| \leq \|x_i^m\|_i$ and $F^i(y_i, x_{\hat{i}}^{m_0}) - F^i(x_i^{m_0}, x_{\hat{i}}^{m_0}) \in intR_+^k$.
Set $y = (y_i, x_{\hat{i}}^{m_0})$, then $y \in X, \|y\| \leq \|x^{m_0}\|$, but

$$\phi(x^{m_0}, y) = F^i(y_i, x_{\hat{i}}^{m_0}) - F^i(x_i^{m_0}, x_{\hat{i}}^{m_0}) \in intR_+^k.$$

Thus, by Theorem 3.3, there exists $x^* \in X$ such that $\phi(x^*, y) \notin intR_+^k$ for any $y \in X$.
$\forall i \in N, \forall y_i \in X_i$, we take $y = (y_i, x_{\hat{i}}^*)$, then $y \in X$.

Then

$$\phi(x^*, y) = \sum_{i=1}^{n} [F_i(y_i, x_{\hat{i}}) - F_i(x_i, x_{\hat{i}})] = F_i(y_i, x_{\hat{i}}^*) - F_i(x_i^*, x_{\hat{i}}^*) \notin intR_+^k,$$

That means $x^* \in X$ is a pseu-weakly Pareto-Nash equilibrium of a multiobjective game.

References

1. Fan, K.: A minimax inequality and applications. In: Shisha, O. (ed.) Inequalities III. Academic Press, New York (1972)
2. Tan, K.K., Yu, J., Yuan, X.Z.: The stability of Ky fan's points. Proc. Am. Math. Soc. **123**(5), 1511–1519 (1995)
3. Hou, S., Gong, X., Yang, X.: Existence and stability of solutions for generalized Ky fan inequality problems with trifunctions. J. Optim. Theor. Appl. **146**(2), 387–398 (2010)
4. Li, X., Wang, Q., Peng, Z.: The stability of set of generalized Ky Fan's points. Positivity **17**(3), 501–513 (2013)
5. Yang, H., Yu, J.: Essential components of the set of weakly Pareto-Nash equilibrium points. Appl. Math. Lett. **15**(5), 553–560 (2002)
6. Yu, J., Peng, D.T.: Solvability of vector Ky fan inequalities with applications. J. Syst. Sci. Complex. **26**(6), 978–990 (2013)
7. Jia, W., Xiang, S., He, J., Yang, Y.L.: The existence and stability for weakly Ky fan's points of set-valued mappings. J. Inequalities Appl. **1**, 1–7 (2012)

8. Kristály, A., Varga, C.: Set-valued versions of Ky Fan's inequality with application to variational inclusion theory. J. Math. Anal. Appl. **282**(1), 8–20 (2003)
9. Qiu, X.L., Peng, D.T.: Some relaxed solutions of minimax inequality for discontinuous game, the 3rd joint China-Dutch workshop and the 7th China meeting, GTA 2016. CCIS **758**, 86–97 (2017)
10. Morgan, J., Scalzo, V.: Pseudocontinuous functions and existence of Nash equilibria. J. Math. Econ. **43**(2), 174–183 (2007)
11. Guide, A.H.: Infinite Dimensional Analysis. Springer Verlag, Berlin (2006). https://doi.org/10.1007/3-540-29587-9
12. Fan, K.: A generalization of Tychonoff's fixed point theorem. Math. Ann. **142**(3), 305–310 (1961)
13. Browder, F.E.: The fixed point theory of multi-valued mappings in topological vector spaces. Math. Ann. **177**(4), 283–301 (1968)
14. Yu, J.: Game Theory Nonlinear Analsis. Science Press, Beijing (2008)

XGBoost-Driven Harsanyi Transformation and Its Application in Incomplete Information Internet Loan Credit Game

Yi-Cheng Gong[1,2(✉)], Yan-Na Zhang[2], and Li Yu[2]

[1] Hubei Province Key Laboratory of Systems Science in Metallurgical Process, Wuhan University of Science and Technology, Wuhan 430065, China
gongyicheng@wust.edu.cn
[2] Department of Mathematics and Statistics, Science College, Wuhan University of Science and Technology, Wuhan 430065, China

Abstract. In theory, the key step of traditional Harsanyi transformation is "*Nature*" assigns types to real players according to certain probability distributions. In big data era, it is still a challenge to obtain the probability distributions in practice, partly because the history type data of a new player is still private. Considering it is easy to access some feature data of a new player as well as to obtain both the feature and type data of massive other players, this paper introduces the statistical learning method eXtreme Gradient Boosting (XGBoost) to propose an XGBoost-driven Harsanyi transformation, where XGBoost is used to predict a new player's type distribution indirectly. To test the effect of XGBoost-driven Harsanyi transformation, an incomplete information Internet loan credit game (3ILCG) is modeled and analyzed. When the loan interest rate $r = 0.2$, the empirical analysis is executed on 24,000 training data and 6,000 testing data. The experiment shows the accuracy (A) and harmonic mean (F_1) of the enterprise loan decision based on p_{xgb} on 6,000 testing data are 0.900833 and 0.945864 respectively. The test experiment demonstrates the XGBoost-driven Harsanyi transformation can help the lending platform to make loan decisions scientifically in practice and improve the practice value of game theory.

Keywords: Game theory · Bayesian Nash equilibrium · Harsanyi transformation · XGBoost · Internet loan

1 Introduction

As an elegant traditional method to analyze the incomplete information game, Harsanyi transformation and then Bayesian Nash equilibrium are introduced by Harsanyi in 1967 [1]. In essence, it transforms uncertainty over the strategy sets of players into uncertainty over their payoffs. By utilizing the transformation, Harsanyi and Selten in 1972 [2] proposed a generalized Nash solution for Two-Person Bargaining Games with Incomplete Information; Harsanyi in 1973 [3] proved players will use their pure strategies approximately with the prescribed probabilities of the mixed strategies equilibrium, due to and the random fluctuations in payoffs. The key step of Harsanyi

D. Li (Ed.): EAGT 2019, CCIS 1082, pp. 116–130, 2019.
https://doi.org/10.1007/978-981-15-0657-4_8

transformation is the virtual player "Nature" assigns types to real players according to certain probability distributions (Harsanyi [3]).

There are many scholars trying to apply the Harsanyi transformation in different fields. For instance, Xiong et al. [4] studied the group decision-making game in 2011, Huang et al. [5] proposed spectrum auction to optimize the allocation of resources through market competition in 2014, Yang et al. [6] analyzed the sequential game in 2015 and Shun et al. [7] constructed the bargaining game model of PPP project risks sharing in 2017. These applications succeeded in combining theoretical Harsanyi Transformation and different incomplete information games. But they did not involve how to get the probability distributions, according to which the virtual player *"Nature"* assigns the real players' types.

Some scholars discussed the methods to obtain the type probability distributions based on the history data of the real players directly (Gong et al. [8] by Mont Carlo 2017)). However, in many real cases, the method will fail when it is difficult to get the history type data of a new player directly.

With the development of big data technology, great progress has been made in data acquisition and processing in 2010s. Some scholars try to develop game theory with the times. Liu and his team [9] in 2013 put forward the concept of "game machine learning" for the first time at the International Artificial Intelligence Congress (IJCAI). Based on online data and Markov chain (Xu et al. [10] in 2013, Li et al. [11] in 2014 and Liu and his team [12] in 2015), some machine learning Advertising Search bidding ranking model is constructed and the practice in Baidu, Tencent and etc. show the game machine learning reduces advertising and increase revenue by 10%. On the other hand, AI games with dynamic scripting are proposed (Laird et al. [13] in 2002; Spronck et al. [14] in 2006, Gibney [15] in 2017). The most famous AI game is about AlphaGo (Silver et al. [16] in 2016), where AlphaGo defeated the human European Go champion 5 to 0 [17]. This is the first time that a computer program has defeated a human professional player in the full-sized game. These achievements inspire that the data can be used to promote the game analysis in practice.

Nowadays, it is easier to get part of a new player's feature data and massive old player's feature data and labeled data directly. In this case, machine learning methods can be used to learn the new player's label and then the type probability distribution. This paper is inspired to get the new player's type distribution indirectly by a new statistical learning method Extreme Gradient Boosting (XGBoost), which is desirable for data of large samples. Based on the predicted distribution, "Nature" assigns the real player's types and the Harsanyi transformation can be driven to be practicable. Because the essence of the idea is data-driven Harsanyi Transformation by the learning method XGBoost, the method will be referred to as XGBoost- Driven Harsanyi Transformation.

2 Preliminary Knowledge

2.1 A Brief Introduction of Traditional Harsanyi Transformation

Generally, an n-player incomplete information game G can be represented as $G=\{A_1,\cdots,A_n;T_1,\cdots,T_n;p_1,\cdots,p_n;u_1,\cdots,u_n\}$. For the ith $(i=1,\cdots,n)$ player, his action a_i belongs to its action space A_i, his type t_i belongs to his type space T_i according to the probability distribution p_i, and u_i is his payoff.

To analyze incomplete information game G, the traditional Harsanyi Transformation is a generally accepted method, which can usually be executed by four steps (Xie [18] in 2017).

Step 1. Introduce a virtual player *"Nature"*, who will select the type $t_{\rm i}$ for the ith $(i = 1, \cdots, n)$ real player randomly before he chooses a strategy $S_i(t_i)$ which is a function of his type $t_i \in T_i$.
Step 2. The *"Nature"* assigns the types for n players $t_1, \cdots, t_{\rm n}$ according to the probability distribution $p_1, \cdots, p_{\rm n}$, where $p_i = p_i(t_{-i}|t_i)$ and $t_{-i} = (t_1, \cdots, t_{i-1}, t_{i+1}, \cdots, t_n)$. The ith player knows his own type t_i and other players' type probability distribution $p_1, \cdots, p_{i-1}, p_{i+1}, \cdots, p_n$.
Step 3. The real players choose actions $a_1, \cdots, a_{\rm n}$ from the action space $A_1, \cdots, A_n$.
Step 4. The real players obtain the payoff $u_i = u_i(a_1, \cdots, a_n, t_i)$.

In essence, the Harsanyi Transformation just introduces the judgment of players' type into the game process. The key problem in its practice is how *"Nature"* get the probability distribution $p_1, \cdots, p_{\rm n}$ to assign the type $t_1, \cdots, t_n$. And this paper tries to employ XGBoost to predict a new player's type based on the related data indirectly.

2.2 A Brief Introduction of XGBoost Model

XGBoost, which is the abbreviation of eXtreme Gradient Boosting, is a machine learning method proposed by Chen Tianqi [19] in 2016. XGBoost aims to promote the speed and accuracy of machine learning method GBDT, which builds a new decision tree based on the negative gradient of empirical loss function, and then prunes the decision tree. The biggest advantage of XGBoost the parallelism by using CPU's multithreading automatically.

For a given data set with n samples and m features $D = \{(x_i, y_i)\}(|D| = {\rm n}, x_i \in R, y_i \in R)$, the final predicted output of XGBoost is the sum of K additive regression tree as (1).

$$\hat{y}(x_i) = \sum\nolimits_{j=1}^{K} f_j(x_i) \tag{1}$$

where $f_j(x_i) = w_q(x_i)\ (q : R^m \rightarrow T, w \in R^T)$ is the predicted output of the jth independent regression trees with leaf weights w (also known as CART), q represents the structure of each tree that maps an example to the corresponding leaf index, T is the number of leaves in the tree.

By adding regular items in the decision tree building stage, XGBoost tries to minimize the objective function shown as formula (2).

$$L(\Phi) = \sum\nolimits_i l(\hat{y}_i, y_i) + \sum\nolimits_k \Omega(f_k) \tag{2}$$

In formula (2), i represents the ith sample, $l(\hat{y}_i, y_i)$ stands for the prediction error of the ith sample, and $\sum_k \Omega(f_k)$ represents the complexity of trees which is shown as formula (3).

$$\Omega(f) = \gamma T + \frac{1}{2}\lambda\|w\|^2 \tag{3}$$

In formula (3), T represents the number of leaf nodes, and w represents the value of the node.

The model in formula (2) includes functions as parameters and cannot be optimized using traditional optimization methods in Euclidean space. It will need to add f_t to minimize the following objective, which is shown in formula (4).

$$L^{(t)} = \sum\nolimits_{i=1}^{n} l\left(y_i, \hat{y}_i^{(t-1)} + f_t(x_i)\right) + \Omega(f_t) \tag{4}$$

To quickly optimize the objective in the general setting, formula (4) is approximated as formula (5) by using the second order Taylor expansion.

$$\tilde{L}^{(t)} \cong \sum\nolimits_{i=1}^{n} [l(y_i, \hat{y}_i^{(t-1)} + g_i f_t(x_i) + \frac{1}{2} h_i f_t^2(x_i)] + \Omega(f_t) \tag{5}$$

In formula (5), $g_i = \partial_{\hat{y}^{(t-1)}} l\left(y_i, \hat{y}^{(t-1)}\right)$, and $h_i = \partial^2_{\hat{y}^{(t-1)}} l\left(y_i, \hat{y}^{(t-1)}\right)$ are the first and second order gradient statistics on the loss function. GBDT uses only first-order derivative information for optimization, while XGBoost uses both first-order and second-order derivatives information, which makes application more extensive.

Being removed the constant terms, formula (5) can be simplified as formula (6).

$$\tilde{L}^{(t)} = \sum\nolimits_{i=1}^{n} \left[g_i f_t(x_i) + \frac{1}{2} h_i f_t^2(x_i) \right] + \Omega(f_t) \tag{6}$$

Define $I_i = \{i | q(x_i) = j\}$ as the instance set of leaf j. By expanding the regularization term Ω, formula (6) can be rewritten as formula (7).

$$\begin{aligned} \tilde{L}^{(t)} &= \sum_{i=1}^{n} \left[g_i f_t(x_i) + \frac{1}{2} h_i f_t^2(x_i) \right] + rT + \frac{1}{2}\lambda \sum_{j=1}^{T} w_j^2 \\ &= \sum\nolimits_{j=1}^{T} [(\sum\nolimits_{i \in I_j} g_i) w_j + \frac{1}{2}\left(\sum\nolimits_{i \in I_j} h_i + \lambda\right) w_j^2] + rT \end{aligned} \tag{7}$$

Let $G_i = \sum_{i \in I_j} g_i$, $H_j = \sum_{i \in I_j} h_i$, then formula (7) can be rewritten as formula (8).

$$\tilde{L}^{(t)} = \sum\nolimits_{j=1}^{T} \left[G_j w_j + \frac{1}{2}(H_j + \lambda) w_j^2 \right] + rT \tag{8}$$

Calculate the partial derivation of formula (8) with respect to w_j, and let it equal to 0, and then the optimal solution w_j^* can be obtained as formula (9).

$$w_j^* = -\frac{G_j}{H_j + \lambda} \tag{9}$$

Then substitute w_j^* into formula (8) to get the optimal objective function as formula (10).

$$\tilde{L}^{(t)} = -\frac{1}{2}\sum_{j=1}^{T}\frac{G_j^2}{H_j + \lambda} + rT \tag{10}$$

According to formula (10), XGBoost can minimize the objective function and realize the high accuracy.

2.3 Evaluation Indicators

To assess the effect of a learning model, some evaluation indicators are needed. In the two classification problem, the model is usually measured by the accuracy (*A*), precision rate (*P*) and recall rate (*R*). To calculate the indicators, it needs to firstly get the confusion matrix, which is shown as Table 1.

Table 1. Confusion matrix

		Real class	
		1	0
Predicted class	1	True positive (*TP*)	False positive (*FP*)
	0	False negative (*FN*)	True negative (*TN*)

Accuracy (*A*) is an indicator to measure the correct proportion of classification, which can be computed by formula (11). Precision rate (*P*) indicates how many real positive samples there are in all predicted positive samples, which can be computed by formula (12). Recall rate (*R*) indicates how many predicted positive samples there are in all positive samples, which can be computed by formula (13). Sometimes it is need to balance the precision rate (*P*) and recall rate (*R*), so the harmonic mean F_1 is defined by formula (14).

$$A = (TP + TN)/(TP + FP + FN + TN) \tag{11}$$

$$P = TP/(TP + FP) \tag{12}$$

$$R = TP/(TP + FN) \tag{13}$$

$$2F_1 = 1/P + 1/R \tag{14}$$

3 An XBGoost-Driving Harsanyi Transformation

As described in the introduction, to execute the Harsanyi Transformation often faces a problem how the *"Nature"* knows the probability distribution tuple $(p_1, \cdots, p_n)$ to assign the type $t = (t_1, \cdots, t_n)$ for players. Given a new player's feature data and massive old player's feature data and labeled data, considering the sample size of data is getting larger and larger, XGBoost is used to learn the new player's label and then the type probability distribution indirectly.

3.1 The Framework of XGBoost-Driven Harsanyi Transformation

To predict the probability distribution of a new player's type scientifically and effectively, the framework of XGBoost-driven Harsanyi transformation can be shown as Fig. 1.

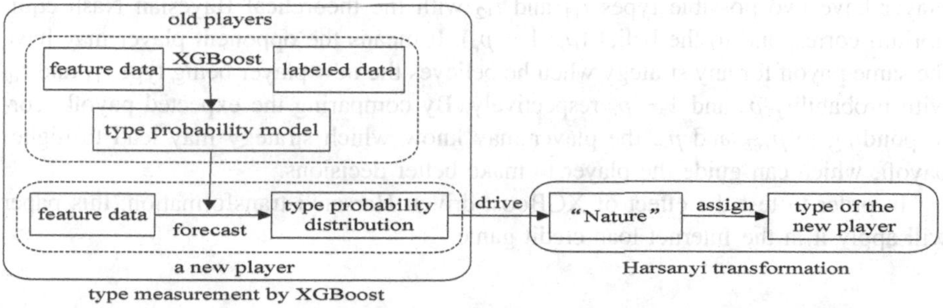

Fig. 1. The framework of XGBoost-driven Harsanyi transformation

Based on massive old players' feature data and labeled data, as shown in Fig. 1, a relationship model between feature data and labeled data can be learnt by XGBoost method. Suppose a new player have feature data without labeled data and two possible types t_{11} and t_{12}, but the type probability distribution is unknown. According to the learnt model, the probability of the new player being of type t_{11}can be predicted as p_{xgb}. And so the probability distribution on the type space $\{t_{11}, t_{12}\}$ is predicted as $(p_{xgb}, 1 - p_{xgb})$, which drives *"Nature"* to assign the new player being of type $t_{11}$with probability p_{xgb} and t_{12} with probability $1 - p_{xgb}$.

Based on this way, XGBoost-driven Harsanyi transformation can help the players in an incomplete information game make decision in practice.

3.2 Game Analysis Based on XGBoost-Driven Harsanyi Transformation

If an agent is in an incomplete information game, the theoretical game analysis based on the traditional Harsanyi transformation can tell him the equilibrium belief and strategy theoretically. To give him a practical reference further, some game analysis based on XGBoost-driven Harsanyi transformation can be done, which is referred to as XGBoost-driven game analysis and shown in Fig. 2.

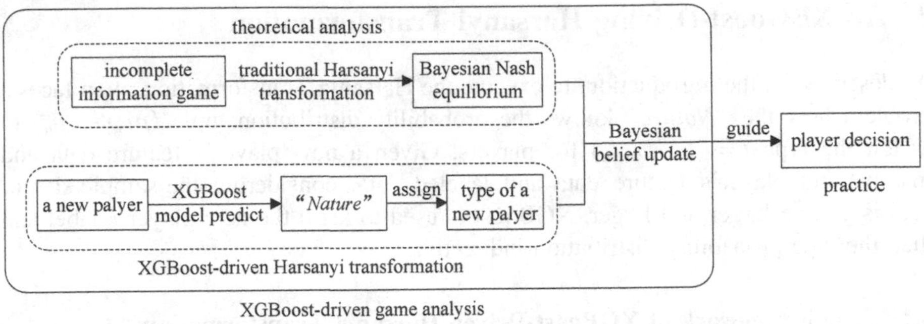

Fig. 2. The framework XGBoost-driven game analysis

As shown in Fig. 2, a player can update his beliefs of his opponent players and be guided to make better decision. For instance, in a two-player game, suppose a new player have two possible types t_{11} and t_{12} with the theoretical Bayesian Nash equilibrium correspond to the belief (p_e, $1 - p_e$). It means the opponent player may have the same payoff for any strategy when he believes the new player being type t_{11} and t_{12} with probability p_e and $1 - p_e$ respectively. By comparing the expected payoffs corresponding to p_{xgb} and p_e, the player may know which strategy may lead to higher payoff, which can guide the player to make better decisions.

In order to test the effect of XGBoost-driven Harsanyi transformation, this paper will apply it in the Internet loan credit game.

4 An Application of XGBoost-Driven Harsanyi Transformation in Incomplete Information Internet Loan Credit Game

4.1 The Incomplete Information Internet Loan Credit Game Model

In the Internet finance loan, there are two kinds of agents, namely the lending platform and applicants. When an applicant applies for a loan, the platform can approve or refuse the application. Because the applicant may hide his own information of credit in order to get a highest payoff, the lending platform needs to analyze the credit of applicant and make a better decision to obtain maximal revenue. Game theory can help the lending platform analyze the situation.

Actually an Internet finance loan can be viewed as a one-player game between the lending platform and applicants, where applicants can take only one strategy *"apply for loan"* while the lending platform can take two strategies *"approve"* or *"refuse"* the application for loan. Because the applicant may have two types, good credit or bad credit which is unknown for the lending platform, the game is an incomplete information game which is referred to as an incomplete information Internet loan credit game (3ILCG). To model the 3ILCG, the following four assumptions are set.

$A1$: The lending platform has only one type: good credit.
$A2$: The amount of the loan applied by the applicant is D.
$A3$: The loan interest rate is r, which is a simple annual interest rate.
$A4$: The opportunity cost is considered in every case.

$A1$ indicates two aspects. Firstly, the type space of the lending platform is a one type set {*good credit*}. Secondly, an applicant believe the lending platform is of *good credit* with probability 1. Namely, the belief of an applicant on the lending platform is absolutely 1. And so the optimal belief of an applicant must be 1. Considering an applicant has only one strategy *"apply for loan"*, the optimal response strategy is only *"apply for loan"*. Therefore the optimal response strategy and belief of an applicant are determined in any Bayesian Nash equilibrium, only the optimal response strategy and belief of lending platform is concerned in the 3ILCG. And so only the payoff of lending platform is concerned in the 3ILCG.

If an applicant is of *"good credit"*, when the lending platform *"approves"* the loan, by $A2$ and $A3$, its payoff is rD; on the other hand, when the lending platform *"refuses"* the loan, by $A4$, and its payoff is $-rD$, because it may lose the opportunity of gaining rD.

If the applicant is of *"bad credit"*, by $A2$ and $A3$, the payoff of lending platform is $-(1 + r)D$ when the lending platform *"approves"* the loan because the lending platform will lose the principal and interest; and by $A4$, the payoff of the lending platform is D when the lending platform *"refuse"* the loan because it will gain the chance of retaining the principal.

Therefore the matrix game model of 3ILCG can be shown as Table 2.

Table 2. Matrix game model of 3ILCG

		applicant	
		apply for loan	
		good credit	*bad credit*
lending platform	*approve*	rD	$-(1 + r)D$
	refuse	$-rD$	D

To analyze the 3ILCG, Harsanyi transformation is an important typical method to transform it into a complete information game.

4.2 The Equilibrium Analysis of Incomplete Information Internet Loan Credit Game

Suppose the virtual player *"Nature"* assigns the applicant's credit type according to the probability distribution $(p_g, 1 - p_g)$ on the type space {*good credit, bad credit*}. By the traditional Harsanyi transformation, 3ILCG can be turned into the extended game model as Fig. 3.

In Fig. 3, the first node is *"Nature"*, who will assign the probability p_g and $1 - p_g$. The number of each terminal indicates the payoff of the lending platform when it reaches the terminal along the corresponding path. By theoretical analysis, Bayesian equilibrium of the 3ILCG can be obtained.

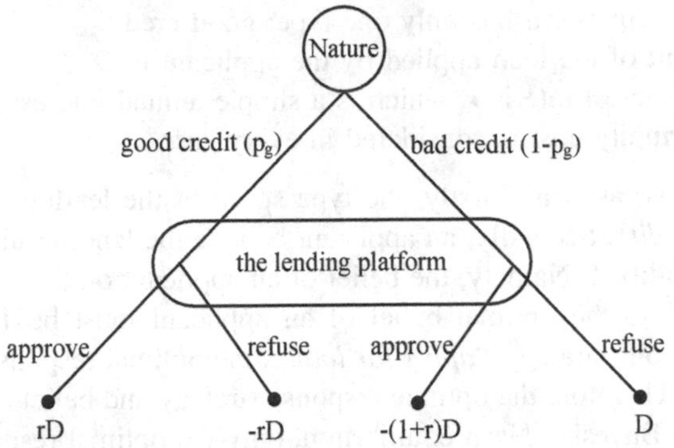

Fig. 3. The extended game of 3ILCG transformed by Harsanyi transformation

When the lending platform approves loan, its payoff is rD or $-(1 + r)D$ with the probability p_g and $1 - p_g$ respectively. So the expected payoff of lending platform taking the strategy *"approve"* is shown as formula (15).

$$E_{approve} = p_g rD + (1 - p_g)[-(1 + r)D] = (2r + 1)p_g D - (1 + r)D \tag{15}$$

Similarly, the expected payoff of lending platform taking the strategy *"refuse"* is shown as formula (16).

$$E_{refuse} = p_g(-rD) + (1 - p_g)D = D - (1 + r)p_g D \tag{16}$$

The equilibrium belief of lending platform about the applicant is the solution of $E_{approve} = E_{refuse}$, that is, $(2r + 1)p_g D - (1 + r)D = D - (1 + r)p_g D$, the Bayesian Nash equilibrium belief p_e of lending platform about the applicant can be obtained as formula (17).

$$p_e = (r + 2)/(3r + 2) \tag{17}$$

If $p_e = p_g$, which is the probability *"Nautre"* assigns a new applicant being of good credit, then $E_{approve} = E_{refuse}$, the expected payoff of lending platform choosing either *"approve"* or *"refuse"* is equal. If $p_e > p_g$, then $E_{approve} < E_{refuse}$, the theoretical expected payoff of the strategy *"approve"* is smaller than the strategy *"refuse"*, so the lending platform would better choose the strategy *"refuse"*. If $p_e < p_g$, then $E_{approve} > E_{refuse}$, the theoretical expected payoff of the strategy *"approve"* is greater than the strategy *"refuse"*, so the lending platform would better choose the strategy *"approve"*.

From formula (17), it can be seen that p_e only depends on loan interest rate r. And its derivative can be calculated as formula (18).

$$P'_e = -4/(3r + 2)^2 \tag{18}$$

Formula (18) shows p_e decreases with r, which means the equilibrium belief p_e is lower when the interest rate r is higher. For the same p_g, the probability of $p_e < p_g$ and lending platform approve the loan increases with r, and so the probability of lending platform suffering the risk of failing in receiving payments increases with r; and vice versa. This conclusion is consistent with loan situation in real word. Therefore, if the lending platform wants to reduce loss, they will choose a reasonable lower loan internet rate r and acquire p_g as accurately as possible; the latter can be helped by XGBoost-driven the Harsanyi transformation.

4.3 XGBoost-Driven Game Analysis of Incomplete Information Internet Loan Credit Game

In the practice of 3ILCG, when a lending platform needs to choose a strategy *"approve"* or *"refuse"*, it is difficult to acquire directly the credit type of a new applicant. However there are massive old lender's data consisting of feature and credit type in its database. Considering all the applicants will fill in a same vacant table for applying a loan, so a new applicant may involve in some same features as the old lenders. Based on the massive old lender' feature and credit type data, a relationship model between feature data and credit type can be learnt by XGBoost method as shown in Fig. 1. And so XGBoost-driven Harsanyi transformation in 3ILCG can be done as Fig. 4.

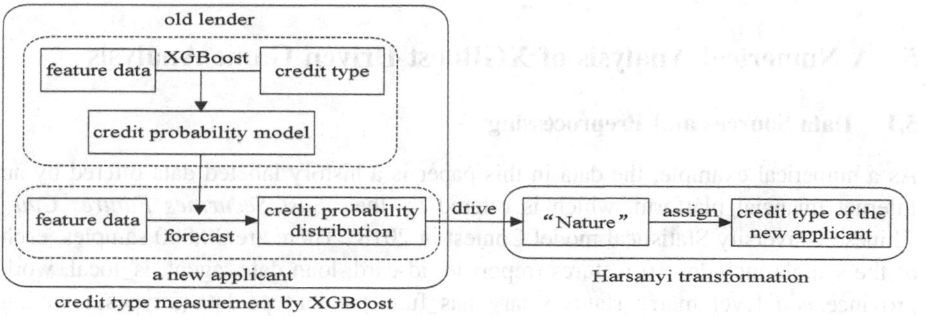

Fig. 4. The framework of XGBoost-driven Harsanyi transformation in 3ILCG

As shown in Fig. 4, inputting the feature data into the learnt credit probability model, the probability of the new applicant being of the type *"good credit"* can be predicted as p_{xgb}, which is a prediction of p_g. And so the credit probability distribution on the type space {*good credit*, *bad credit*} $(p_g, 1 - p_g)$ is predicted as $(p_{xgb}, 1 - p_{xgb})$, which drives *"Nature"* to assign the new player being of type *"good credit"* with probability p_{xgb} and *bad credit* with probability $1 - p_{xgb}$. And the lending platform can observe $(p_{xgb}, 1 - p_{xgb})$.

To help the lending platform take better strategy, XGBoost-driven game analysis in 3ILCG can be executed as shown in Fig. 5.

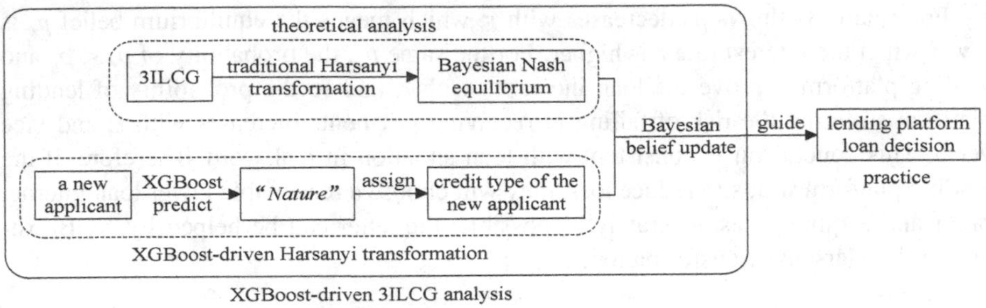

Fig. 5. The framework of XGBoost-driven game analysis in 3ILCG

As shown in Fig. 5, the lending platform can update his belief by comparing p_{xgb} and the theoretical equilibrium belief p_e. Besides being guided theoretically by Bayesian Nash equilibrium strategy and belief p_e, lending platform can observe "*Nature*" assigns the new player being of type "*good credit*" or "*bad credit*" with probability and p_{xgb} and $1 - p_{xgb}$. As a prediction of the probability p_g of a new applicant being of "*good credit*", p_{xgb} is between 0 and 1 and a bigger p_{xgb} means the probability of applicant being of "*good credit*" is higher. And so the lending platform can update his belief and strategy in this way: if $p_{xgb} \geqslant p_e$, the lending platform will "*approve*" the loan of applicant in order to increase its income; if $p_{xgb} < p_e$, the lending platform will "*refuse*" the loan of applicant in order to reduce its loss.

5 A Numerical Analysis of XGBoost-Driven Game Analysis

5.1 Data Sources and Preprocessing

As a numerical example, the data in this paper is a history labeled data offered by an Internet financial platform, which is offered by the *"East Securities Futures Cup"* Chinese University Statistical model Contest in 2018,. There are 30,000 samples, each of the sample includes 10 features (report id, id-card, loan date, agent, is_local, work province, edu_level, marry_status, salary, has_fund.) and a type data (2 types, repaying on time or not).

Before learning a model, the data are preprocessed as following 2 aspects.

(1) Delete the first three features, because they are trivial.

After deleting the first three features, each sample includes 7 features (agent, is_local, work province, edu_level, marry_status, salary, has_fund.), which are named $x_1, \ldots, x_7$ sequentially.

(2) Quantity 5 qualitative features by one-hot code.

For an instance, the feature *edu_level* is mainly divided into four categories and quantified as integers from 0 to 3, where 0 means unclear or unfilled information, 1 means college degrees or below, 2 means undergraduate degrees, and 3 means master or above. And then by one hot encode, 0 to 3 can be respectively coded into 1000, 0100, 0010 and 0001. These quantitative data can be used to train an XGBoost model.

5.2 Numerical Simulation

As the internet loan interest rate is generally around 20%, the annual interest rate is set as $r = 20\%$ in this numerical example. Substitute $r = 20\%$ into formula (17) in Sect. 4.2, equilibrium strategy can be calculated as formula (19) and partly shown in Column 3 of Table 3.

$$p_e = (r+2)/(3r+2) = (0.2+2)/(3 \times 0.2+2) = 0.8462 \quad (19)$$

Table 3. The partial results of a new applicant learned by XGBoost on the test set

ID	p_{xgb}	p_e	Strategy of the lending platforms
1	0.368999762	0.8462	*refuse*
2	0.873951237	0.8462	*approve*
…	…	…	…
6000	0.653426285	0.8462	*refuse*

Before learning an XGBoost model, the 30000 data is divided into a training set and a test set at simple random division with the proportion of training set is 80%. The training set consisting of 24000 samples is used to learn an XGBoost model as in Figs. 6 and 7.

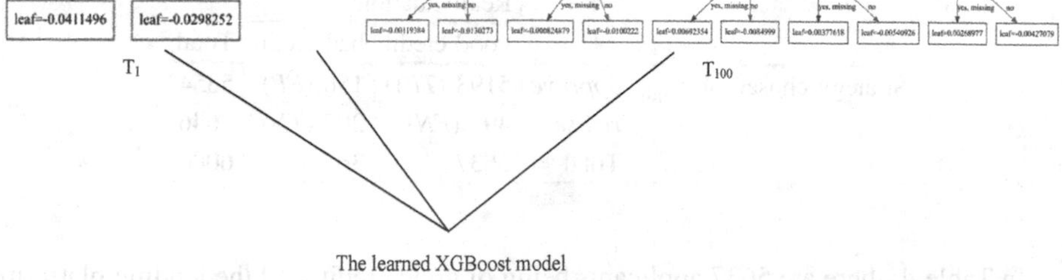

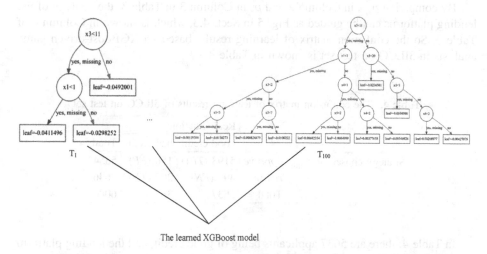

Fig. 6. The out sketch of The learned XGBoost model consisting of 100 trees

From Fig. 6, we observed that the XGBoost model is constituted of 100 decision trees. Some trees are small and others are big, which indicates the adaptive weights of samples according to the previous tree's error. On the other hand, in the first tree T_1 of Fig. 6, it can be seen the first two important features are x_3 (is_local) and x_1 (agent).

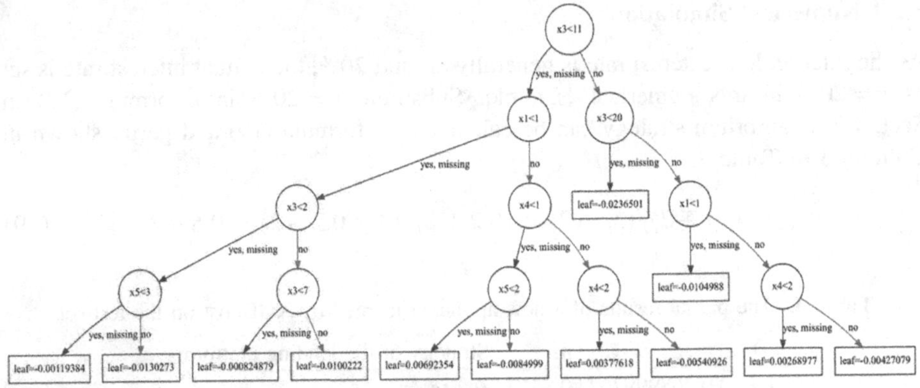

Fig. 7. The last tree T_{100} of learned XGBoost model consisting of 100 trees

Figure 7 is the clear figure of the last tree T_{100} in Fig. 6 it indicates there are four important features, which are x_3, x_1, x_5 (marry_status) *and* x_4 (edu_level). So Fig. 6 shows in different trees, the importance of features may differ. Therefore different trees have different error and loss, the accuracy of the whole learned XGBoost model as formula (1) is better than a single tree.

The testing set consisting of 6000 samples is used to test the learned XGBoost in Figs. 6 and 7, p_{xgb} of each applicant on the test set a can be predicted, and some results are shown in Column 2 of Table 3.

By comparing p_{xgb} in Column 2 and p_e in Column 3 in Table 3, the strategy of the lending platforms can be guided as Fig. 5 in Sect. 4.3, which is shown in Column 4 of Table 3. So the confusion matrix of learning results based on XGBoost-driven game analysis in 3ILCG on test set is shown in Table 4.

Table 4. The confusion matrix of learning results of 3ILCG on test set

		Real situation		
		good credit	bad credit	Total
Strategy chosen on p_{xgb}	*approve*	5198 (*TP*)	156 (*FP*)	5354
	refuse	493 (*FN*)	207 (*TN*)	646
	Total	5637	363	6000

In Table 4, there are 5637 applicants being of good credit, and the lending platform is guided by XGBoost-driven Harsanyi transformation to choose the strategy *"approve"* for 5198 applicants and *"refuse"* for 493 applicants. In this case, because the strategy *"approve"* can lead to the greater payoff than the strategy *"refuse"* for the lending platform, the learning result *"approve"* is the right strategy which corresponds to *TP* in Table 1 in Sect. 2.3, while the learning result *"refuse"* is the wrong strategy which corresponds to *FN* in Table 1 in Sect. 2.3.

On the other hand, there are 363 applicants being of bad credit, and the lending platform is guided by XGBoost-driven Harsanyi transformation to choose the strategy *"approve"* for 156 applicants and *"refuse"* for 207 applicants. In this case, because the strategy *"refuse"* can lead to the smaller loss than the strategy *"approve"* for the lending platform, the learning result *"refuse"* is the right strategy which corresponds to *TN* in Table 1 in Sect. 2.3, which corresponds to *FP* in Table 1 in Sect. 2.3, while the learning result

According to Table 4 and the evaluation indicators in Sect. 2.3, the evaluation indicators of lending platform decisions of 3ILCG based on XGBoost-driven Harsanyi transformation can be calculated as formulas (20)–(23).

$$A = (5198 + 207)/6000 = 0.900833 \tag{20}$$

$$P = 5198/5354 = 0.970863 \tag{21}$$

$$R = 5198/5637 = 0.922122 \tag{22}$$

$$F_1 = 0.945864 \tag{23}$$

Obviously, the accuracy (A) 0.900833 and harmonic mean (F_1) 0.945864 are both reasonably high. The results of empirical analysis show that 3ILCG analysis based on XGBoost-driven Harsanyi transformation has a great performance.

6 Conclusion

This paper combines the traditional game theory and XGBoost. The traditional Harsanyi Transformation often faces the difficulty in getting type distribution of the players, according to which the *"Nature"* assigns the types. This paper proposes an XGBoost-driven Harsanyi transformation method to employ the type distributions, which are predicted indirectly based on the relevant feature data and type data by the statistical learning method XGBoost. In the empirical analysis, XGBoost-dri*ven* Harsanyi transformation model is specifically applied to the Internet finance loan, which is a 3ILCG model. A numerical example of 3ILCG analysis based on XGBoost-driven Harsanyi transformation is done on 24000 training data and 6000 test data, the accuracy (A) and harmonic mean (F_1) are respectively 0.900833 and 0.945864. The empirical analysis means 3ILCG based on XGBoost-driv*en* Harsanyi transformation can help the lending platform to make loan decisions scientifically in practice.

XGBoost-driv*en* Harsanyi transformation can improve the practice value of game theory and augment the application field of the statistical learning method XGBoost. The hybrid researches of game theory and statistical learning methods are expecting.

Acknowledge. Thanks to the support from National Natural Science Foundation of China (Project Number: 61671338).

References

1. Harsanyi, J.C.: Games with incomplete information played by "Bayesian" players, I-III Part I, the basic model. Manag. Sci. INFORMS **14**(3), 159–182 (1967)
2. Harsanyi, J.C., Selten, R.: A generalized Nash solution for two-person bargaining games with incomplete information. Manag. Sci. **18**, 80–106 (1972)
3. Harsanyi, J.C.: Games with randomly disturbed payoffs: a new rationale for mixed-strategy equilibrium points. Int. J. Game Theory **2**(1), 1–23 (1973)
4. Xiong, F., Liu, Y., Si, X., Chen, H.: Group decision-making simulation with incomplete information. Syst. Eng. Theory Pract. **31**(1), 151–157 (2011)
5. Huang, H., Sun, Y.E., Chen, Z.L., Xu, H.L., Xing, K., Chen, G.L.: A study on complete competitive equilibrium of spectrum two-way auction mechanism. Comput. Res. Dev. **51**(3), 479–490 (2014)
6. Yang, Y., Zhu, W., Zhang, X.: New mechanism design in the C2C online reputation evaluation optimizing. In: Proceedings of the 13th International Conference on Enterprise Information Systems, pp. 1021–1029. Springer, Berlin (2015)
7. Shun, Z., Mingshun, L.I.: Game analysis of PPP project risk sharing under the condition of incomplete information. Eng. Econ. **27**(4), 37–41 (2017)
8. Gong, Y., Zhang, Y., Zhao, J., Yu, L. Pan, T.: Incomplete information game and simulation of logistics supervision. In: 2017 2nd International Conference on Applied Mathematics, Simulation and Modeling, pp. 284–289. Atlantis Press, Bangkok (2017)
9. He, D., Chen, W., Wang, L., Liu, T.Y.: A Game-theoretic machine learning approach for revenue maximization in, sponsored search. In: Proceedings of the Twenty-Third International Joint Conference on Artificial Intelligence, pp. 284–289. The AAAI Press, Palo Alto (2013)
10. Xu, H., Gao, B., Yang, D., Liu, T.Y.: Predicting advertiser bidding behaviors in sponsored search by rationality modeling. In: Proceedings of the 22nd International conference on World Wide Web, pp. 176–186. Association for Computing Machinery Press, New York (2013)
11. Li, H., Tian, F., Chen, W., Qin, T., Liu, T. Y: Generalization analysis for game- theoretic machine learning. In: Proceedings of the Twenty-Ninth AAAI Conference on Artificial Intelligence, pp. 2089–2095. The AAAI Press, Palo Alto (2014)
12. Liu, T.Y., Chen, W., Qin, T.: Mechanism learning with mechanism induced data. In: Twenty-ninth AAAI Conference on Artificial Intelligence, pp. 4037–4041. The AAAI Press, Palo Alto (2015)
13. Laird, J.E.: Research in human-level AI using computer games. Commun. ACM **45**(1), 32–35 (2002)
14. Spronck, P., Ponsen, M., Sprinkhuizen-Kuyper, I., Postma, E.: Adaptive game AI with dynamic scripting. Mach. Learn. **63**(3), 217–248 (2006)
15. Gibney, E.: Self-taught AI is best yet at strategy game Go. Nature **10**(1), 68–74 (2017)
16. Silver, D., Huang, A., Maddison, C.J., Guez, A., Hassabis, D.: Mastering the game of go with deep neural networks and tree search. Nature **529**(7587), 484–489 (2016)
17. Elizabeth, G.: Google AI algorithm masters ancient game of Go. Nature **529**(7587), 445–446 (2016)
18. Xie, Z.: Economic Game Theory, 4th edn. Fudan University Press, Shanghai (2002)
19. Chen, T., Guestrin, C.: Xgboost: a scalable tree boosting system. In: Proceedings of the 22nd ACM SIGKDD International Conference on Knowledge Discovery and Data Mining. Association for Computing Machinery Press, New York (2016)

The Method for Solving Bi-matrix Games with Intuitionistic Fuzzy Set Payoffs

Jiang-Xia Nan[1], Li Zhang[1], and Deng-Feng Li[2(✉)]

[1] School of Mathematics and Computing Science, Guangxi Colleges and Universities Key Laboratory of Data Analysis and Computation, Guilin University of Electronic Technology, Guilin 541004, Guangxi, China
jiangxia1107@163.com

[2] School of Economics and Management, Fuzhou University, Fuzhou 350108, Fujian, China
lidengfeng@fzu.edu.cn

Abstract. The aim of this paper is to develop a bilinear programming method for solving bi-matrix games in which the payoffs are expressed with intuitionistic fuzzy sets (IFSs), which are called IFS bi-matrix games for short. In this method, using the equivalent relation between IFSs and interval-valued fuzzy sets (IVFSs) and the operations of IVFSs, we propose a new order relation of IFSs through introducing a ranking function, which is proven to be a total order relation. Hereby we introduce the concepts of solutions of IFS bi-matrix games and parametric bi-matrix games. It is proven that any IFS bi-matrix game has at least one satisfying Nash equilibrium solution, which is equivalent to the Nash equilibrium solution of corresponding parametric bi-matrix game. The latter can be obtained through solving the auxiliary parametric bilinear programming model. The models and method proposed in this paper are demonstrated with a real example of the e-commerce retailers' strategy choice problem.

Keywords: Noncooperative game · Intuitionistic fuzzy set · Bilinear programming · Fuzzy game

1 Introduction

Bi-matrix games, which are an important type of two-person nonzero-sum noncooperative games in normal-form, have been successfully applied to different areas such as finance, management, economics and business. The bi-matrix games assume that the payoffs are represented with crisp values, which indicate that the payoffs are exactly known by players. However, players often are not able to evaluate exactly the payoffs due to a lack of information and/or imprecision of the available information in real game situations. In order to make bi-matrix game theory more applicable to real competitive decision problems, the fuzzy set introduced by Zadeh [1] has been used to describe imprecise and uncertain information appearing in real game problems. Hence, fuzzy bi-matrix games have been extensively studied. Using the ranking method of fuzzy numbers, Vijay et al. [2] studied the bi-matrix game with fuzzy goals and fuzzy

D. Li (Ed.): EAGT 2019, CCIS 1082, pp. 131–150, 2019.
https://doi.org/10.1007/978-981-15-0657-4_9

payoffs. Using the possibility measure of fuzzy numbers, Meada [3] introduced two concepts of equilibrium for the bi-matrix games with fuzzy payoffs. Bector and Chandra [4] studied bi-matrix games with fuzzy payoffs and fuzzy goals based on some duality of fuzzy linear programming. Larbani [5] proposed an approach to solve fuzzy bi-matrix games based on the idea of introducing "Nature" as a player in fuzzy multi-attribute decision making problems.

The fuzzy set uses only a membership function, which assigns to each element x of the universal set a real number $\mu(x) \in [0, 1]$ to indicate the degree of belongingness to the fuzzy set under consideration. The degree of nonbelongingness is just automatically equal to $1 - \mu(x)$. However, people who express the degree of membership of a given element in a fuzzy set very often do not express corresponding degree of nonmembership as the complement to 1. Sometimes it seems to be more natural to describe imprecise and uncertain opinions not only by membership functions. It is due to the fact that in some situations it is easier to describe our negative feelings than the positive attitudes. Even more, quite often people can easily specify objects or alternatives they dislike. For example, it may happen that customers are asked about their satisfaction degrees for electronic commerce (e-commerce) retailers A, B and C. Often customers may not determine which is more satisfactory for e-commerce retailers A and B, but they feel sure that they dissatisfy with e-commerce retailer C. This case is suitably modeled by the intuitionistic fuzzy set (IFS), which was firstly introduced by Atanassov [6]. The IFS is characterized by two functions which are expressed the degree of belongingness and the degree of nonbelongingness, respectively. The IFS may make people consider independently positive and negative information and has been applied to some fields (Aggarwal et al. [13]; Li [9, 11, 23]; Li and Cheng [8]; Li and Nan [10]; Nan et al. [12]). It is expected to apply the IFS to dealing with imprecise information in bi-matrix game problems. As far as we know, however, there exists little investigation on bi-matrix games using the IFS. Nayak and Pal [14] defined the concept of a Nash equilibrium solution for bi-matrix games with intuitionistic fuzzy goals and discussed existence of the Nash equilibrium solutions based on the inclusion relation of IFSs. In some game situations, the payoffs are often imprecise. Thus, it is necessary to study bi-matrix games with IFS payoffs, which are often called IFS bi-matrix games for short. Obviously, IFS bi-matrix games remarkably differ from fuzzy bi-matrix games since the former uses both membership and nonmembership degrees to express the payoffs while the latter only uses membership degrees to express the payoffs. As a result, the fuzzy bi-matrix game models and methods can not be directly used to solve IFS bi-matrix games. Therefore, the aim of this paper is to formulate IFS bi-matrix games and propose their solutions, concepts and corresponding solving methods.

The rest of this paper is organized as follows. Section 2 briefly introduces the concepts of IFSs and bi-matrix games and establishes a new order relation of IFSs. Section 3 formulates IFS bi-matrix games and proposes corresponding solving method based on the constructed auxiliary parametric bilinear programming model, which is derived from the defined order relation of IFSs and the bi-matrix game model. In Sect. 4, the models and method proposed in this paper are illustrated with a real example of the e-commerce retailers' strategy choice problem. Conclusion is given in Sect. 5.

2 Definitions and Notations

2.1 The Definition and Operations of IFSs

Definition 1 (Atanassov [6, 7]). Let $X = \{x_1, x_2, \cdots, x_L\}$ be a finite universal set. An IFS $\tilde{C}$ in X is an object expressed as $\tilde{C} = \{< x_l, \mu_{\tilde{C}}(x_l), \upsilon_{\tilde{C}}(x_l) > | x_l \in X\}$, where $\mu_{\tilde{C}}(x_l) \in [0, 1]$ and $\upsilon_{\tilde{C}}(x_l) \in [0, 1]$ are respectively the membership and nonmembership degrees of an element $x_l \in X$ to the set $\tilde{C} \subseteq X$ so that they satisfy the condition: $0 \le \mu_{\tilde{C}}(x_l) + \upsilon_{\tilde{C}}(x_l) \le 1$ for every $x_l \in X$.

Let

$$\pi_{\tilde{C}}(x_l) = 1 - \mu_{\tilde{C}}(x_l) - \upsilon_{\tilde{C}}(x_l), \tag{1}$$

which is called the intuitionistic index of an element x_l in the IFS $\tilde{C}$. Obviously, $0 \le \pi_{\tilde{C}}(x_l) \le 1$.

For example, $\tilde{C}_0 = \{<x_1, 0.4, 0.5>, <x_2, 0.5, 0.3>, <x_3, 0.6, 0>, <x_4, 0, 0.7>\}$ is an IFS in the finite universal set $X_1 = \{x_1, x_2, x_3, x_4\}$. Sometimes, the IFS $\tilde{C}_0$ may be written as follows:

$$\tilde{C}_0 = <0.4, 0.5> / x_1 + <0.5, 0.3> / x_2 + <0.6, 0> / x_3 + <0, 0.7> / x_4$$

or

$$\tilde{C}_0 = \frac{<0.4, 0.5>}{x_1} + \frac{<0.5, 0.3>}{x_2} + \frac{<0.6, 0>}{x_3} + \frac{<0, 0.7>}{x_4}.$$

Definition 2 (Atanassov [6, 7]). Let $\tilde{A} = \{< x_l, \mu_{\tilde{A}}(x_l), \upsilon_{\tilde{A}}(x_l) > | x_l \in X\}$ and $\tilde{B} = \{< x_l, \mu_{\tilde{B}}(x_l), \upsilon_{\tilde{B}}(x_l) > | x_l \in X\}$ be two IFSs in the finite universal set $X = \{x_1, x_2, \cdots, x_L\}$ and $\gamma > 0$ be a real number. The operations over IFSs are stipulated as follows:

(1) $\tilde{A} + \tilde{B} = \{< x_l, \mu_{\tilde{A}}(x_l) + \mu_{\tilde{B}}(x_l) - \mu_{\tilde{A}}(x_l)\mu_{\tilde{B}}(x_l), \upsilon_{\tilde{A}}(x_l)\upsilon_{\tilde{B}}(x_l) > | x_l \in X\}$;
(2) $\tilde{A}\tilde{B} = \{< x_l, \mu_{\tilde{A}}(x_l)\mu_{\tilde{B}}(x_l), \upsilon_{\tilde{A}}(x_l) + \upsilon_{\tilde{B}}(x_l) - \upsilon_{\tilde{A}}(x_l)\upsilon_{\tilde{B}}(x_l) > | x_l \in X\}$;
(3) $\gamma\tilde{A} = \{< x_l,\ 1 - (1 - \mu_{\tilde{A}}(x_l))^\gamma,\ (\upsilon_{\tilde{A}}(x_l))^\gamma > | x_l \in X\}$.

2.2 The New Ranking Method of IFSs and Properties

When IFSs are used to model game problems, their comparison or ranking order is important (Li [15, 16]; Zhang [17]). It is noticed that an IFS is mathematically equivalent to an interval-valued fuzzy set (IVFS). In order to facilitate the sequent discussions, inspired by the order relation of IVFSs, a simple ranking method (or ranking function) of IFSs is defined as follows.

The set of IFSs $\tilde{C}(x_l) = \{<x_l, \mu_{\tilde{C}}(x_l), \upsilon_{\tilde{C}}(x_l)>\}$ is denoted by $\Re(X)$. An IFS $\tilde{C}(x_l) = \{<x_l, \mu_{\tilde{C}}(x_l), \upsilon_{\tilde{C}}(x_l)>\}$ is mathematically equivalent to an IVFS $[\mu_{\tilde{C}}(x_l), 1-\upsilon_{\tilde{C}}(x_l)]$ (or $[\mu_{\tilde{C}}(x_l), \mu_{\tilde{C}}(x_l)+\pi_{\tilde{C}}(x_l)]$).

Remark 1. According to the notation introduced by Yager [22], the above IFS $\tilde{C}(x_l) = \{<x_l, \mu_{\tilde{C}}(x_l), \upsilon_{\tilde{C}}(x_l)>\}$ can be concisely expressed with the tuple $\tilde{C}(x_l) = <\mu_{\tilde{C}}(x_l), \upsilon_{\tilde{C}}(x_l)>$, which is called the intuitionistic membership grade (or degree).

Let $F_\lambda : \Re(X) \to [0,1]$ be a function so that

$$F_\lambda(\tilde{C}(x_l)) = F_\lambda([\mu_{\tilde{C}}(x_l), 1-\upsilon_{\tilde{C}}(x_l)]) = \mu_{\tilde{C}}(x_l) + \lambda(1-\upsilon_{\tilde{C}}(x_l) - \mu_{\tilde{C}}(x_l)), \quad (2)$$

where $\tilde{C} \in \Re(X)$ and $\lambda \in [0,1]$. The parameter λ represents a weight which reflects players' preference information about hesitancy. The larger λ the more the hesitancy is assigned as part of the membership degree. That is to say, the larger value of λ indicates that we favor alternatives with more imprecise membership degree, i.e., the larger λ tends to be more optimistic with respect to the allocation of the intuitionistic index. On the other hand, the smaller λ shows that hesitancy gets more resolved in favor of nonmembership degree. It tends to resolve uncertainty in favor of nonmembership degree, i.e., the smaller λ tends to resolve the uncertainty in membership degree in a more pessimistic way.

For any given $\lambda \in [0,1]$, using the ranking function F_λ (i.e., Eq. (2)), players with different preference information can compare the IFSs in the set $\Re(X)$. Thus, we can define the order relations "$\leq_{IFS}$" and "$\geq_{IFS}$" on the set $\Re(X)$ as follows.

Definition 3. Let $\tilde{A}(x_l) = \{<x_l, \mu_{\tilde{A}}(x_l), \upsilon_{\tilde{A}}(x_l)>\}$ and $\tilde{B}(x_l) = \{<x_l, \mu_{\tilde{B}}(x_l), \upsilon_{\tilde{B}}(x_l)>\}$ be any IFSs in the set $\Re(X)$. Then, $\tilde{A}(x_l) \leq_{IFS} \tilde{B}(x_l)$ if and only if for a given parameter $\lambda \in [0,1]$, $F_\lambda(\tilde{A}(x_l)) \leq F_\lambda(\tilde{B}(x_l))$.

The symbol "$\leq_{IFS}$" is an intuitionistic fuzzy version of the order relation "$\leq$" in the real number set and has the linguistic interpretation "essentially smaller than or equal to".

Similarly, the order relation "$\geq_{IFS}$" can be defined. Namely, $\tilde{A}(x_l) \geq_{IFS} \tilde{B}(x_l)$ if and only if for a given parameter $\lambda \in [0,1]$, $F_\lambda(\tilde{A}(x_l)) \geq F_\lambda(\tilde{B}(x_l))$.

Theorem 1. For a given $\lambda \in [0,1]$, the order relations "$\leq_{IFS}$" and "$\geq_{IFS}$" are a total order on the set $\Re(X)$.

Proof. We only prove the order relation "$\leq_{IFS}$" is a total order on the set $\Re(X)$. The relation "$\geq_{IFS}$" can be proven in the same way.

(1) It is obviously that $\tilde{A}(x_l) \leq_{IFS} \tilde{A}(x_l)$ for any IFS $\tilde{A}(x_l) \in \Re(X)$ since $F_\lambda(\tilde{A}(x_l)) \leq F_\lambda(\tilde{A}(x_l))$ is always valid according to Eq. (2). Thus, the order relation "$\leq_{IFS}$" satisfies reflexivity.
(2) For a given parameter $\lambda \in [0,1]$, according to Definition 3, we have $F_\lambda(\tilde{A}(x_l)) \leq F_\lambda(\tilde{B}(x_l))$ and $F_\lambda(\tilde{B}(x_l)) \leq F_\lambda(\tilde{A}(x_l))$ if $\tilde{A}(x_l) \leq_{IFS} \tilde{B}(x_l)$ and $\tilde{B}(x_l) \leq_{IFS} \tilde{A}(x_l)$. Hence, $F_\lambda(\tilde{A}(x_l)) = F_\lambda(\tilde{B}(x_l))$. Therefore, we have $\tilde{A}(x_l) =_{IFS} \tilde{B}(x_l)$. Thus, the order relation "$\leq_{IFS}$" satisfies antisymmetry.

(3) Let $\tilde{A}(x_l)$, $\tilde{B}(x_l)$ and $\tilde{C}(x_l)$ be IFSs in the set $\Re(X)$. For a given $\lambda \in [0,1]$, according to Definition 3, we have both $F_\lambda(\tilde{A}(x_l)) \le F_\lambda(\tilde{B}(x_l))$ and $F_\lambda(\tilde{B}(x_l)) \le F_\lambda(\tilde{C}(x_l))$ if $\tilde{A}(x_l) \le_{IFS} \tilde{B}(x_l)$ and $\tilde{B}(x_l) \le_{IFS} \tilde{C}(x_l)$. Hence, $F_\lambda(\tilde{A}(x_l)) \le F_\lambda(\tilde{C}(x_l))$. It directly follows from Definition 3 that $\tilde{A}(x_l) \le_{IFS} \tilde{C}(x_l)$. Thus, the order relation "$\le_{IFS}$" satisfies transitivity.

It can be easily seen from the above cases (1)–(3) that the order relation "$\le_{IFS}$" is a partial order on the set $\Re(X)$. Furthermore, for any two IFSs $\tilde{A}(x_l)$ and $\tilde{B}(x_l)$ in the set $\Re(X)$, it can always ensure that either $F_\lambda(\tilde{A}(x_l)) \le F_\lambda(\tilde{B}(x_l))$ or $F_\lambda(\tilde{B}(x_l)) \le F_\lambda(\tilde{A}(x_l))$ according to Eq. (2). Hereby it directly follows that either $\tilde{A}(x_l) \le_{IFS} \tilde{B}(x_l)$ or $\tilde{B}(x_l) \le_{IFS} \tilde{A}(x_l)$ according to Definition 3. Thus, the order relation "$\le_{IFS}$" is a total order on $\Re(X)$. ∎

It can be easily seen that the ranking function F_λ have some useful properties, which are summarized in Theorems 2 and 3.

Theorem 2. Assume that $\tilde{C}(x_l) = \{<x_l, \mu_{\tilde{C}}(x_l), \upsilon_{\tilde{C}}(x_l)>\}$ is any IFS in the set $\Re(X)$. Then, the following equation is always valid:

$$F_\lambda(\delta\tilde{C}(x_l)) = \delta F_\lambda(\tilde{C}(x_l)). \tag{3}$$

Proof. It follows from Definition 3 and the operations of IVFSs that

$$\begin{aligned} F_\lambda(\delta\tilde{C}(x_l)) &= F_\lambda(\delta[\mu_{\tilde{C}}(x_l), 1-\upsilon_{\tilde{C}}(x_l)]) \\ &= F_\lambda([\delta\mu_{\tilde{C}}(x_l), \delta(1-\upsilon_{\tilde{C}}(x_l))]) \\ &= \delta\mu_{\tilde{C}}(x_l) + \lambda[\delta(1-\upsilon_{\tilde{C}}(x_l)) - \delta\mu_{\tilde{C}}(x_l)] \\ &= \delta[\mu_{\tilde{C}}(x_l) + \lambda(1-\upsilon_{\tilde{C}}(x_l) - \mu_{\tilde{C}}(x_l))] \\ &= \delta F_\lambda(\tilde{C}(x_l)), \end{aligned}$$

i.e., $F_\lambda(\delta\tilde{C}(x_l)) = \delta F_\lambda(\tilde{C}(x_l))$. Thus, we have completed the Proof of Theorem 2. ∎

Theorem 3. Assume that $\tilde{A}(x_l) = \{<x_l, \mu_{\tilde{A}}(x_l), \upsilon_{\tilde{A}}(x_l)>\}$ and $\tilde{B}(x_l) = \{<x_l, \mu_{\tilde{B}}(x_l), \upsilon_{\tilde{B}}(x_l)>\}$ are any IFSs in the set $\Re(X)$. Then, the following equation is valid:

$$F_\lambda(\tilde{A}(x_l) + \tilde{B}(x_l)) = F_\lambda(\tilde{A}(x_l)) + F_\lambda(\tilde{B}(x_l)). \tag{4}$$

Proof. According to Definition 3 and the operations of IVFSs, we have

$$\begin{aligned} F_\lambda(\tilde{A}(x_l) + \tilde{B}(x_l)) &= F_\lambda([\mu_{\tilde{A}}(x_l), 1-\upsilon_{\tilde{A}}(x_l)] + [\mu_{\tilde{B}}(x_l), 1-\upsilon_{\tilde{B}}(x_l)]) \\ &= F_\lambda([\mu_{\tilde{A}}(x_l) + \mu_{\tilde{B}}(x_l), 1-\upsilon_{\tilde{A}}(x_l) + 1 - \upsilon_{\tilde{B}}(x_l)]) \\ &= (\mu_{\tilde{A}}(x_l) + \mu_{\tilde{B}}(x_l)) + \lambda[(1-\upsilon_{\tilde{A}}(x_l) + 1 - \upsilon_{\tilde{B}}(x_l)) - (\mu_{\tilde{A}}(x_l) + \mu_{\tilde{B}}(x_l))] \\ &= [\mu_{\tilde{A}}(x_l) + \lambda(1-\upsilon_{\tilde{A}}(x_l) - \mu_{\tilde{A}}(x_l))] + [\mu_{\tilde{B}}(x_l) + \lambda(1-\upsilon_{\tilde{B}}(x_l) - \mu_{\tilde{B}}(x_l))] \\ &= F_\lambda(\tilde{A}(x_l)) + F_\lambda(\tilde{B}(x_l)). \end{aligned}$$

i.e., $F_\lambda(\tilde{A}(x_l) + \tilde{B}(x_l)) = F_\lambda(\tilde{A}(x_l)) + F_\lambda(\tilde{B}(x_l))$. Thus, we have proven Theorem 3. ∎

The above Theorems 2 and 3 show that the ranking function F_λ is linear.

2.3 Bi-matrix Games and Auxiliary Bilinear Programming Models

Assume that $S_1 = \{\alpha_1, \alpha_2, \cdots, \alpha_m\}$ and $S_2 = \{\beta_1, \beta_2, \cdots, \beta_n\}$ are sets of pure strategies for players I and II, respectively. The payoff matrices of players I and II are expressed with $\boldsymbol{A} = (a_{ij})_{m\times n}$ and $\boldsymbol{B} = (b_{ij})_{m\times n}$, respectively. The vectors $\boldsymbol{y} = (y_1, y_2, \cdots, y_m)^{\mathrm{T}}$ and $\boldsymbol{z} = (z_1, z_2, \cdots, z_n)^{\mathrm{T}}$ are mixed strategies for players I and II, where y_i $(i = 1, 2, \cdots, m)$ and z_j $(j = 1, 2, \cdots, n)$ are probabilities in which players I and II choose their pure strategies $\alpha_i \in S_1 (i = 1, 2, \cdots, m)$ and $\beta_j \in S_2 (j = 1, 2, \cdots, n)$, respectively; the symbol "T" is the transpose of a vector/matrix. Sets of all mixed strategies for I and II are denoted by Y and Z, i.e., $Y = \{\boldsymbol{y} | \sum_{i=1}^{m} y_i = 1, y_i \geq 0\ (i = 1, 2, \cdots, m)\}$ and $Z = \{\boldsymbol{z} | \sum_{j=1}^{n} z_j = 1, z_j \geq 0\ (j = 1, 2, \cdots, n)\}$, respectively. Thus, a two-person nonzero-sum finite game may be expressed with the triple $(Y, Z, \boldsymbol{A}, \boldsymbol{B})$. In the sequent, such a game usually is simply called the bi-matrix game $(\boldsymbol{A}, \boldsymbol{B})$ in which both players want to maximize his/her own payoffs.

Suppose that players I and II are playing the bi-matrix game $(\boldsymbol{A}, \boldsymbol{B})$. When I chooses any mixed strategy $\boldsymbol{y} \in Y$ and II chooses any mixed strategy $\boldsymbol{z} \in Z$, then the expected payoffs of I and II can be computed as $E_1(\boldsymbol{y}, \boldsymbol{z}) = \boldsymbol{y}^{\mathrm{T}} \boldsymbol{A} \boldsymbol{z} = \sum_{i=1}^{m} \sum_{j=1}^{n} y_i a_{ij} z_j$ and $E_2(\boldsymbol{y}, \boldsymbol{z}) = \boldsymbol{y}^{\mathrm{T}} \boldsymbol{B} \boldsymbol{z} = \sum_{i=1}^{m} \sum_{j=1}^{n} y_i b_{ij} z_j$, respectively.

Definition 4 (Owen [18]). If there is a pair $(\boldsymbol{y}^*, \boldsymbol{z}^*) \in Y \times Z$ so that

$$\boldsymbol{y}^{\mathrm{T}} \boldsymbol{A} \boldsymbol{z}^* \leq \boldsymbol{y}^{*\mathrm{T}} \boldsymbol{A} \boldsymbol{z}^* \text{ for any } \boldsymbol{y} \in Y$$

and

$$\boldsymbol{y}^{*\mathrm{T}} \boldsymbol{B} \boldsymbol{z} \leq \boldsymbol{y}^{*\mathrm{T}} \boldsymbol{B} \boldsymbol{z}^* \text{ for any } \boldsymbol{z} \in Z,$$

then $(\boldsymbol{y}^*, \boldsymbol{z}^*)$ is called a Nash equilibrium point of the bi-matrix game $(\boldsymbol{A}, \boldsymbol{B})$, $\boldsymbol{y}^*$ and $\boldsymbol{z}^*$ are called Nash equilibrium strategies of players I and II, $u^* = \boldsymbol{y}^{*\mathrm{T}} \boldsymbol{A} \boldsymbol{z}^*$ and $v^* = \boldsymbol{y}^{*\mathrm{T}} \boldsymbol{B} \boldsymbol{z}^*$ are called Nash equilibrium values of players I and II, respectively. $(\boldsymbol{y}^{*\mathrm{T}}, \boldsymbol{z}^{*\mathrm{T}}, u^*, v^*)$ is called a Nash equilibrium solution of the bi-matrix game $(\boldsymbol{A}, \boldsymbol{B})$.

The following theorem guarantees existence of Nash equilibrium solutions of any bi-matrix game.

Theorem 4 (Nash [19]). Any bi-matrix game $(\boldsymbol{A}, \boldsymbol{B})$ has at least one Nash equilibrium solution.

A Nash equilibrium solution of any bi-matrix game $(\boldsymbol{A}, \boldsymbol{B})$ can be obtained by solving the bilinear programming model stated as the following Theorem 5.

Theorem 5 (Mangasarian and Stone [20]). Let $(\boldsymbol{A},\boldsymbol{B})$ be any bi-matrix game. $(\boldsymbol{y}^{*\mathrm{T}},\boldsymbol{z}^{*\mathrm{T}},u^*,v^*)$ is a Nash equilibrium solution of the bi-matrix game $(\boldsymbol{A},\boldsymbol{B})$ if and only if it is a solution of the bilinear programming model as follows:

$$\max\{\boldsymbol{y}^{\mathrm{T}}(\boldsymbol{A}+\boldsymbol{B})\boldsymbol{z}-u-v\}$$
$$s.t.\begin{cases}\boldsymbol{A}\boldsymbol{z}\leq u\boldsymbol{e}^m\\ \boldsymbol{B}^{\mathrm{T}}\boldsymbol{y}\leq v\boldsymbol{e}^n\\ \boldsymbol{y}^{\mathrm{T}}\boldsymbol{e}^m=1\\ \boldsymbol{z}^{\mathrm{T}}\boldsymbol{e}^n=1\\ \boldsymbol{y}\geq\boldsymbol{0},\boldsymbol{z}\geq\boldsymbol{0}\end{cases}\tag{5}$$

Furthermore, if $(\boldsymbol{y}^{*\mathrm{T}},\boldsymbol{z}^{*\mathrm{T}},u^*,v^*)$ is a solution of the above bilinear programming model, then $u^*=\boldsymbol{y}^{*\mathrm{T}}\boldsymbol{A}\boldsymbol{z}^*$, $v^*=\boldsymbol{y}^{*\mathrm{T}}\boldsymbol{B}\boldsymbol{z}^*$ and $\boldsymbol{y}^{*\mathrm{T}}(\boldsymbol{A}+\boldsymbol{B})\boldsymbol{z}^*-u^*-v^*=0$.

Equation (5) can be expressed in the vector format as the following bilinear programming model:

$$\max\{\sum_{j=1}^{n}\sum_{i=1}^{m}y_i(a_{ij}+b_{ij})z_j-u-v\}$$
$$s.t.\begin{cases}\sum\limits_{j=1}^{n}a_{ij}z_j\leq u\quad(i=1,2,\cdots,m)\\ \sum\limits_{i=1}^{m}b_{ij}y_i\leq v\quad(j=1,2,\cdots,n)\\ y_1+y_2+\cdots+y_m=1\\ z_1+z_2+\cdots+z_n=1\\ y_i\geq 0\quad(i=1,2,\cdots,m)\\ z_j\geq 0\quad(j=1,2,\cdots,n)\end{cases}\tag{6}$$

3 Models and Method for IFS Bi-matrix Games

Let us consider an IFS bi-matrix game, where sets of pure strategies S_1 and S_2 and sets of mixed strategies Y and Z for players I and II are defined as the above Sect. 2.3. If player I chooses any pure strategy $\alpha_i\in S_1$ $(i=1,2,\cdots,m)$ and player II chooses any pure strategy $\beta_j\in S_2$ $(j=1,2,\cdots,n)$, then at the situation (α_i,β_j) players I and II gain payoffs, which are expressed with IFSs $\tilde{A}_{ij}(\alpha_i,\beta_j)=\{<(\alpha_i,\beta_j),\mu_{\tilde{A}_{ij}},\upsilon_{\tilde{A}_{ij}}>\}$ and $\tilde{B}_{ij}(\alpha_i,\beta_j)=\{<(\alpha_i,\beta_j),\mu_{\tilde{B}_{ij}},\upsilon_{\tilde{B}_{ij}}>\}$ $(i=1,2,\cdots,m;\ j=1,2,\cdots,n)$, respectively. Thus, the payoff matrices of players I and II are expressed as $\tilde{\boldsymbol{A}}=(\tilde{A}_{ij}(\alpha_i,\beta_j))_{m\times n}$ and $\tilde{\boldsymbol{B}}=(\tilde{B}_{ij}(\alpha_i,\beta_j))_{m\times n}$, respectively.

Remark 2. As stated earlier, the above payoffs $\tilde{A}_{ij}(\alpha_i, \beta_j) = \{<(\alpha_i, \beta_j), \mu_{\tilde{A}_{ij}}, \upsilon_{\tilde{A}_{ij}}>\}$ and $\tilde{B}_{ij}(\alpha_i, \beta_j) = \{<(\alpha_i, \beta_j), \mu_{\tilde{B}_{ij}}, \upsilon_{\tilde{B}_{ij}}>\}$ $(i = 1, 2, \cdots, m; j = 1, 2, \cdots, n)$ of players I and II are IFSs in the finite universal set

$$X' = \{(\alpha_i, \beta_j) \mid \alpha_i \in S_1 \ (i = 1, 2, \cdots, m), \beta_j \in S_2 \ (j = 1, 2, \cdots, n)\}.$$

For instance, let us consider a simple example in which there are two pure strategies for both players I and II, i.e., player I has pure strategies α_1 and α_2 and player II has pure strategies β_1 and β_2. Then, the universal set is $X'_0 = \{(\alpha_1, \beta_1), (\alpha_1, \beta_2), (\alpha_2, \beta_1), (\alpha_2, \beta_2)\}$, which has four elements (i.e., situations). Payoffs of players I and II at the above situation (α_1, β_1) are expressed with the IFSs $\tilde{A}_{11}(\alpha_1, \beta_1) = \{<(\alpha_1, \beta_1), 0.7, 0.2>\}$ and $\tilde{B}_{11}(\alpha_1, \beta_1) = \{<(\alpha_1, \beta_1), 0.3, 0.4>\}$, respectively. According to the above notation of the IFS and convention, $\tilde{A}_{11}(\alpha_1, \beta_1)$ may be written as $\tilde{A}_{11} = <0.7, 0.2>/(\alpha_1, \beta_1)$ or $\tilde{A}_{11} = \frac{<0.7,0.2>}{(\alpha_1,\beta_1)}$. According to the notation introdued by Yager (2009), $\tilde{A}_{11}(\alpha_1, \beta_1)$ is concisely written as $\tilde{A}_{11}(\alpha_1, \beta_1) = <0.7, 0.2>$. I-n the same way, $\tilde{B}_{11}(\alpha_1, \beta_1)$ can be explained.

In the sequel, the above IFS bi-matrix game is simply denoted by $(\tilde{\boldsymbol{A}}, \tilde{\boldsymbol{B}})$ for short. It is customary to assume that players I and II want to maximize his/her own IFS payoffs.

If player I chooses any mixed strategy $\boldsymbol{y} \in Y$ and player II chooses any mixed strategy $\boldsymbol{z} \in Z$, then the expected payoff of player I is defined as follows:

$$\tilde{E}_1(\boldsymbol{y}, \boldsymbol{z}) = \boldsymbol{y}^{\mathrm{T}} \tilde{\boldsymbol{A}} \boldsymbol{z}. \tag{7}$$

According to Definition 2, the expected payoff $\tilde{E}_1(\boldsymbol{y}, \boldsymbol{z})$ of player I is an IFS and can be calculated as follows:

$$\tilde{E}_1(\boldsymbol{y}, \boldsymbol{z}) = \{<(\bar{y}, \bar{z}), 1 - \prod_{j=1}^{n} [\prod_{i=1}^{m} (1 - \mu_{\tilde{A}_{ij}})^{y_i z_j}], \prod_{j=1}^{n} (\prod_{i=1}^{m} \upsilon_{\tilde{A}_{ij}}^{y_i z_j}) >\} \tag{8}$$

where $(\bar{y}, \bar{z})$ represents a mixed situation, which corresponds to the mixed strategies $\boldsymbol{y}$ and $\boldsymbol{z}$.

Similarly, the expected payoff of player II is defined as follows:

$$\tilde{E}_2(\boldsymbol{y}, \boldsymbol{z}) = \boldsymbol{y}^{\mathrm{T}} \tilde{\boldsymbol{B}} \boldsymbol{z}, \tag{9}$$

which is also an IFS and can be obtained as follows:

$$\tilde{E}_2(\boldsymbol{y}, \boldsymbol{z}) = \{<(\bar{y}', \bar{z}'), 1 - \prod_{j=1}^{n} [\prod_{i=1}^{m} (1 - \mu_{\tilde{B}_{ij}})^{y_i z_j}], \prod_{j=1}^{n} (\prod_{i=1}^{m} \upsilon_{\tilde{B}_{ij}}^{y_i z_j}) >\}, \tag{10}$$

where $(\bar{y}', \bar{z}')$ represents a mixed situation, which corresponds to the mixed strategies $\boldsymbol{y}$ and $\boldsymbol{z}$.

Definition 5. Assume that there is a pair $(y^*, z^*) \in Y \times Z$. If any $y \in Y$ and $z \in Z$ satisfy

$$y^{\mathrm{T}}\tilde{A}z^* \leq_{IFS} y^{*\mathrm{T}}\tilde{A}z^*$$

and

$$y^{*\mathrm{T}}\tilde{B}z \leq_{IFS} y^{*\mathrm{T}}\tilde{B}z^*,$$

then (y^*, z^*) is called a Nash equilibrium point of the IFS bi-matrix game $(\tilde{A}, \tilde{B})$, y^* and z^* are called Nash equilibrium strategies of players I and II, $\tilde{u}^* = y^{*\mathrm{T}}\tilde{A}z^*$ and $\tilde{v}^* = y^{*\mathrm{T}}\tilde{B}z^*$ are called Nash equilibrium values of players I and II, respectively. $(y^*, z^*, \tilde{u}^*, \tilde{v}^*)$ is called a Nash equilibrium solution of the IFS bi-matrix game $(\tilde{A}, \tilde{B})$.

Stated as earlier, however, player I's expected payoff $y^{\mathrm{T}}\tilde{A}z$ and player II's expected payoff $y^{\mathrm{T}}\tilde{B}z$ are IFSs. Therefore, there are no commonly-used concepts of solutions of IFS bi-matrix games. Furthermore, it is not easy to compute the membership degrees and the nonmembership degrees of players' expected payoffs. As a result, solving Nash equilibrium solutions of IFS bi-matrix games are very difficult. In the sequel, we use the ranking function F_λ (i.e., Eq. (2)) to develop a new method for solving the IFS bi-matrix game $(\tilde{A}, \tilde{B})$.

Using the ranking function of IFSs given by Eq. (2), we can transform the IFS payoff matrices $\tilde{A}$ and $\tilde{B}$ of players I and II into the payoff matrices as follows:

$$A(\lambda_1) = F_{\lambda_1}((\tilde{A}_{ij}(\alpha_i, \beta_j))_{m\times n}) = (F_{\lambda_1}(\tilde{A}_{ij}(\alpha_i, \beta_j)))_{m\times n} \tag{11}$$

and

$$B(\lambda_2) = F_{\lambda_2}((\tilde{B}_{ij}(\alpha_i, \beta_j))_{m\times n}) = (F_{\lambda_2}(\tilde{B}_{ij}(\alpha_i, \beta_j)))_{m\times n} \tag{12}$$

respectively, where

$$F_{\lambda_1}(\tilde{A}_{ij}(\alpha_i, \beta_j)) = \mu_{\tilde{A}_{ij}(\alpha_i,\beta_j)} + \lambda_1(1 - \mu_{\tilde{A}_{ij}(\alpha_i,\beta_j)} - \upsilon_{\tilde{A}_{ij}(\alpha_i,\beta_j)}), \lambda_1 \in [0,1]\lambda_2 \in [0,1]$$

and

$$F_{\lambda_2}(\tilde{B}_{ij}(\alpha_i, \beta_j)) = \mu_{\tilde{B}_{ij}(\alpha_i,\beta_j)} + \lambda_2(1 - \mu_{\tilde{B}_{ij}(\alpha_i,\beta_j)} - \upsilon_{\tilde{B}_{ij}(\alpha_i,\beta_j)})\ (i = 1, 2, \cdots, m; j = 1, 2, \cdots, n).$$

It can be easily seen that Eqs. (11) and (12) may reflect players,preference information. $\lambda_1 \in [0, 1/2)$ (or $\lambda_2 \in [0, 1/2)$) shows that player I (or II) is risk-averse. $\lambda_1 \in (1/2, 1]$ (or $\lambda_2 \in (1/2, 1]$) shows that player I (or II) is risk-prone. $\lambda_1 = 1/2$ (or $\lambda_2 = 1/2$) shows that player I (or II) is neutral.

According to the above usage and notations, the above parametric bi-matrix game can be simply denoted by $(A(\lambda_1), B(\lambda_2))$, where the pure (or mixed) strategy sets of players I and II are S_1 and S_2 (or Y and Z) defined as the above Sect. 2.3, respectively.

Then, the IFS bi-matrix game $(\tilde{\boldsymbol{A}}, \tilde{\boldsymbol{B}})$ is transformed into the parametric bi-matrix game $(\boldsymbol{A}(\lambda_1), \boldsymbol{B}(\lambda_2))$. Hereby, according to Definitions 3–5 and Theorems 2 and 3, we can give the definition of satisfying Nash equilibrium solutions of the IFS bi-matrix game $(\tilde{\boldsymbol{A}}, \tilde{\boldsymbol{B}})$ as follows.

Definition 6. For given parameters $\lambda_1 \in [0, 1]$ and $\lambda_2 \in [0, 1]$, if there is a pair $(\boldsymbol{y}^*, \boldsymbol{z}^*) \in Y \times Z$ so that any $\boldsymbol{y} \in Y$ and $\boldsymbol{z} \in Z$ satisfy the following conditions:

$$\boldsymbol{y}^{\mathrm{T}} F_{\lambda_1}(\tilde{\boldsymbol{A}}) \boldsymbol{z}^* \leq \boldsymbol{y}^{*\mathrm{T}} F_{\lambda_1}(\tilde{\boldsymbol{A}}) \boldsymbol{z}^*$$

and

$$\boldsymbol{y}^{*\mathrm{T}} F_{\lambda_2}(\tilde{\boldsymbol{B}}) \boldsymbol{z} \leq \boldsymbol{y}^{*\mathrm{T}} F_{\lambda_2}(\tilde{\boldsymbol{B}}) \boldsymbol{z}^*,$$

then $(\boldsymbol{y}^*, \boldsymbol{z}^*)$ is called a satisfying Nash equilibrium point of the IFS bi-matrix game $(\tilde{\boldsymbol{A}}, \tilde{\boldsymbol{B}})$, $\boldsymbol{y}^*$ and $\boldsymbol{z}^*$ are called satisfying Nash equilibrium strategies of players I and II, $u^*(\lambda_1) = \boldsymbol{y}^{*\mathrm{T}} F_{\lambda_1}(\tilde{\boldsymbol{A}}) \boldsymbol{z}^*$ and $v^*(\lambda_2) = \boldsymbol{y}^{*\mathrm{T}} F_{\lambda_2}(\tilde{\boldsymbol{B}}) \boldsymbol{z}^*$ are called satisfying equilibrium values of players I and II, respectively. $(\boldsymbol{y}^*, \boldsymbol{z}^*, u^*(\lambda_1), v^*(\lambda_2))$ is called a satisfying Nash equilibrium solution of the IFS bi-matrix game $(\tilde{\boldsymbol{A}}, \tilde{\boldsymbol{B}})$.

It can be easily seen from the ranking function given by Eq. (2) and Theorems 2 and 3 that Definitions 5 and 6 are equivalent in the sense of the order relation "$\leq_{IFS}$" defined by Definition 3.

Thus, for given parameters $\lambda_1 \in [0, 1]$ and $\lambda_2 \in [0, 1]$, according to Theorem 4 (Nash [19]), the parametric bi-matrix game $(\boldsymbol{A}(\lambda_1), \boldsymbol{B}(\lambda_2))$ has at least one Nash equilibrium solution. Namely, the IFS bi-matrix game $(\tilde{\boldsymbol{A}}, \tilde{\boldsymbol{B}})$ has at least one satisfying Nash equilibrium solution, which can be obtained through solving the following parametric bilinear programming model according to Theorem 5 (i.e., Eq. (6) or Eq. (5)):

$$\max \left\{ \begin{array}{l} \sum_{j=1}^{n} \sum_{i=1}^{m} y_i [\mu_{\tilde{A}_{ij}(\alpha_i,\beta_j)} + \lambda_1 (1 - \mu_{\tilde{A}_{ij}(\alpha_i,\beta_j)} - \upsilon_{\tilde{A}_{ij}(\alpha_i,\beta_j)}) + \mu_{\tilde{B}_{ij}(\alpha_i,\beta_j)} \\ + \lambda_2 (1 - \mu_{\tilde{B}_{ij}(\alpha_i,\beta_j)} - \upsilon_{\tilde{B}_{ij}(\alpha_i,\beta_j)})] z_j - u(\lambda_1) - v(\lambda_2) \end{array} \right\}$$

$$s.t. \begin{cases} \sum_{j=1}^{n} [\mu_{\tilde{A}_{ij}(\alpha_i,\beta_j)} + \lambda_1 (1 - \mu_{\tilde{A}_{ij}(\alpha_i,\beta_j)} - \upsilon_{\tilde{A}_{ij}(\alpha_i,\beta_j)})] z_j \leq u(\lambda_1) \ (i = 1, 2, \cdots, m) \\ \sum_{i=1}^{m} [\mu_{\tilde{B}_{ij}(\alpha_i,\beta_j)} + \lambda_2 (1 - \mu_{\tilde{B}_{ij}(\alpha_i,\beta_j)} - \upsilon_{\tilde{B}_{ij}(\alpha_i,\beta_j)})] y_i \leq v(\lambda_2) \ (j = 1, 2, \cdots, n) \\ y_1 + y_2 + \cdots + y_m = 1 \\ z_1 + z_2 + \cdots + z_n = 1 \\ v(\lambda_2) \geq 0, u(\lambda_1) \geq 0 \\ y_i \geq 0 \ (i = 1, 2, \cdots, m), z_j \geq 0 \ (j = 1, 2, \cdots, n), \end{cases} \tag{13}$$

where y_i $(i = 1, 2, \cdots, m)$, z_j $(j = 1, 2, \cdots, n)$, $u(\lambda_1)$ and $v(\lambda_2)$ are decision variables.

According to Theorem 5, if $(\boldsymbol{y}^*, \boldsymbol{z}^*, u^*(\lambda_1), v^*(\lambda_2))$ is a solution of the above parametric bilinear programming model (i.e., Eq. (13)), then

$$u^*(\lambda_1) = \boldsymbol{y}^{*\mathrm{T}} F_{\lambda_1}(\tilde{\boldsymbol{A}})\boldsymbol{z}^* = \sum_{j=1}^{n}\sum_{i=1}^{m}[\mu_{\tilde{A}_{ij}(\alpha_i,\beta_j)} + \lambda_1(1 - \mu_{\tilde{A}_{ij}(\alpha_i,\beta_j)} - \upsilon_{\tilde{A}_{ij}(\alpha_i,\beta_j)})]y_i^* z_j^*,$$

$$v^*(\lambda_2) = \boldsymbol{y}^{*\mathrm{T}} F_{\lambda_2}(\tilde{\boldsymbol{B}})\boldsymbol{z}^* = \sum_{j=1}^{n}\sum_{i=1}^{m}[\mu_{\tilde{B}_{ij}(\alpha_i,\beta_j)} + \lambda_2(1 - \mu_{\tilde{B}_{ij}(\alpha_i,\beta_j)} - \upsilon_{\tilde{B}_{ij}(\alpha_i,\beta_j)})]y_i^* z_j^*,$$

and

$$\boldsymbol{y}^{*\mathrm{T}}(F_{\lambda_1}(\tilde{\boldsymbol{A}}) + F_{\lambda_2}(\tilde{\boldsymbol{B}}))\boldsymbol{z}^* \text{ - } u^*(\lambda_1) \text{ - } v^*(\lambda_2) = 0.$$

Noticing that $y_i^* \geq 0$, $z_j^* \geq 0$, $\mu_{\tilde{A}_{ij}(\alpha_i,\beta_j)} + \upsilon_{\tilde{A}_{ij}(\alpha_i,\beta_j)} \leq 1$ and $\mu_{\tilde{B}_{ij}(\alpha_i,\beta_j)} + \upsilon_{\tilde{B}_{ij}(\alpha_i,\beta_j)} \leq 1$, it follows that

$$\frac{\partial u^*(\lambda_1)}{\partial \lambda_1} = \sum_{j=1}^{n}\sum_{i=1}^{m}(1 - \mu_{\tilde{A}_{ij}(\alpha_i,\beta_j)} - \upsilon_{\tilde{A}_{ij}(\alpha_i,\beta_j)})y_i^* z_j^* \geq 0$$

and

$$\frac{\partial v^*(\lambda_2)}{\partial \lambda_2} = \sum_{j=1}^{n}\sum_{i=1}^{m}(1 - \mu_{\tilde{B}_{ij}(\alpha_i,\beta_j)} - \upsilon_{\tilde{B}_{ij}(\alpha_i,\beta_j)})y_i^* z_j^* \geq 0,$$

respectively. Then, $u^*(\lambda_1)$ and $v^*(\lambda_2)$ are monotonic and nondecreasing functions of the parameters $\lambda_1 \in [0, 1]$ and $\lambda_2 \in [0, 1]$, respectively. Thus, the satisfying Nash equilibrium values of players I and II are obtained as the IVFSs (i.e., intervals) $[u^*(0), u^*(1)]$ and $[v^*(0), v^*(1)]$, respectively. According to the equivalent relation between IFSs and IVFSs, the IVFSs $[u^*(0), u^*(1)]$ and $[v^*(0), v^*(1)]$ can be written as the IFSs

$$\{< (\bar{y}^*, \bar{z}^*), u^*(0), 1 - u^*(1) > \}$$

and

$$\{< (\bar{y}^*, \bar{z}^*), v^*(0), 1 - v^*(1) > \},$$

respectively, denote by

$$\tilde{u}^*(\bar{y}^*, \bar{z}^*) = \{< (\bar{y}^*, \bar{z}^*), \mu_{\tilde{u}^*(\bar{y}^*,\bar{z}^*)}, \upsilon_{\tilde{u}^*(\bar{y}^*,\bar{z}^*)} > \} = \{< (\bar{y}^*, \bar{z}^*), u^*(0), 1 - u^*(1) > \}$$

and

$$\tilde{v}^*(\bar{y}^*, \bar{z}^*) = \{< (\bar{y}^*, \bar{z}^*), \mu_{\tilde{v}^*(\bar{y}^*,\bar{z}^*)}, \upsilon_{\tilde{v}^*(\bar{y}^*,\bar{z}^*)} > \} = \{< (\bar{y}^*, \bar{z}^*), v^*(0), 1 - v^*(1) > \},$$

where $(\bar{y}^*, \bar{z}^*)$ represents a mixed situation. Then, the membership degrees of the satisfying Nash equilibrium values $\tilde{u}^*(\bar{y}^*, \bar{z}^*)$ and $\tilde{v}^*(\bar{y}^*, \bar{z}^*)$ of players I and II are $\mu_{\tilde{u}^*(\bar{y}^*,\bar{z}^*)} = u^*(0)$ and $\mu_{\tilde{v}^*(\bar{y}^*,\bar{z}^*)} = v^*(0)$, and the nonmembership degrees of the satisfying Nash equilibrium values $\tilde{u}^*(\bar{y}^*, \bar{z}^*)$ and $\tilde{v}^*(\bar{y}^*, \bar{z}^*)$ of players I and II are $\upsilon_{\tilde{u}^*(\bar{y}^*,\bar{z}^*)} = 1 - u^*(1)$ and $\upsilon_{\tilde{v}^*(\bar{y}^*,\bar{z}^*)} = 1 - v^*(1)$, respectively.

In particular, for the parameters $\lambda_1 = 0$ and $\lambda_2 = 0$, Eq. (13) becomes the bilinear-programming model as follows:

$$\max\{\sum_{j=1}^{n}\sum_{i=1}^{m} y_i(\mu_{\tilde{A}_{ij}(\alpha_i,\beta_j)} + \mu_{\tilde{B}_{ij}(\alpha_i,\beta_j)})z_j - \mu_{\tilde{u}(\bar{x},\bar{y})} - \mu_{\tilde{v}(\bar{x},\bar{y})}\}$$

$$s.t.\begin{cases} \sum_{j=1}^{n} \mu_{\tilde{A}_{ij}(\alpha_i,\beta_j)} z_j \le \mu_{\tilde{u}(\bar{x},\bar{y})} \ \ (i = 1, 2, \cdots, m) \\ \sum_{i=1}^{m} \mu_{\tilde{B}_{ij}(\alpha_i,\beta_j)} y_i \le \mu_{\tilde{v}(\bar{x},\bar{y})} \ \ (j = 1, 2, \cdots, n) \\ y_1 + y_2 + \cdots + y_m = 1 \\ z_1 + z_2 + \cdots + z_n = 1 \\ 0 \le \mu_{\tilde{u}(\bar{x},\bar{y})} \le 1 \\ 0 \le \mu_{\tilde{v}(\bar{x},\bar{y})} \le 1 \\ y_i \ge 0 \ \ (i = 1, 2, \cdots, m) \\ z_j \ge 0 \ \ (j = 1, 2, \cdots, n), \end{cases} \tag{14}$$

where y_i $(i = 1, 2, \cdots, m)$, z_j $(j = 1, 2, \cdots, n)$, $\mu_{\tilde{u}(\bar{x},\bar{y})}$ and $\mu_{\tilde{v}(\bar{x},\bar{y})}$ are decision variables. The solution of the above bilinear programming model (i.e., Eq. (14)) can be obtained by the Lemke-Howson's algorithm (Lemke and Howson [21]), denoted by $(y^{*T}, z^{*T}, \mu_{\tilde{u}^*(\bar{x}^*,\bar{y}^*)}, \mu_{\tilde{v}^*(\bar{x}^*,\bar{y}^*)})$.

Similarly, for the parameters $\lambda_1 = 1$ and $\lambda_2 = 1$, Eq. (13) becomes the bilinear-programming model as follows:

$$\max\{\sum_{j=1}^{n}\sum_{i=1}^{m} y_i(2 - \upsilon_{\tilde{A}_{ij}(\alpha_i,\beta_j)} - \upsilon_{\tilde{B}_{ij}(\alpha_i,\beta_j)})z_j - (2 - \upsilon_{\tilde{u}(\bar{x},\bar{y})} - \upsilon_{\tilde{v}(\bar{x},\bar{y})})\}$$

$$s.t.\begin{cases} \sum_{j=1}^{n} (1 - \upsilon_{\tilde{A}_{ij}(\alpha_i,\beta_j)}) z_j \le 1 - \upsilon_{\tilde{u}(\bar{x},\bar{y})} \ \ (i = 1, 2, \cdots, m) \\ \sum_{i=1}^{m} (1 - \upsilon_{\tilde{B}_{ij}(\alpha_i,\beta_j)}) y_i \le 1 - \upsilon_{\tilde{v}(\bar{x},\bar{y})} \ \ (j = 1, 2, \cdots, n) \\ y_1 + y_2 + \cdots + y_m = 1 \\ z_1 + z_2 + \cdots + z_n = 1 \\ 0 \le \upsilon_{\tilde{v}(\bar{x},\bar{y})} \le 1 \\ 0 \le \upsilon_{\tilde{u}(\bar{x},\bar{y})} \le 1 \\ y_i \ge 0 \ \ (i = 1, 2, \cdots, m) \\ z_j \ge 0 \ \ (j = 1, 2, \cdots, n), \end{cases} \tag{15}$$

where y_i $(i = 1, 2, \cdots, m)$, z_j $(j = 1, 2, \cdots, n)$, $\upsilon_{\tilde{u}(\bar{x},\bar{y})}$ and $\upsilon_{\tilde{v}(\bar{x},\bar{y})}$ are decision variables. Likewise, the solution of Eq. (15) can be obtained by the Lemke-Howson's algorithm (Lemke and Howson, [21]), denoted by $(\boldsymbol{y}'^{*\mathrm{T}}, \boldsymbol{z}'^{*\mathrm{T}}, \upsilon_{\tilde{u}'^{*}(\bar{x}'^{*},\bar{y}'^{*})}, \upsilon_{\tilde{v}'^{*}(\bar{x}'^{*},\bar{y}'^{*})})$.

Thus, we can explicitly obtain the satisfying Nash equilibrium values and corresponding satisfying Nash equilibrium strategies of players I and II through solving the derived two bilinear programming models (i.e., Eqs. (14) and (15)). Furthermore, according to Eq. (13), any satisfying Nash equilibrium values and corresponding satisfying Nash equilibrium strategies of players I and II can be obtained through choosing different parameters $\lambda_1 \in [0, 1]$ and $\lambda_2 \in [0, 1]$.

4 An Example of an E-Commerce Retailers' Strategy Choice Problem

There are many applications of the classical game theory to real decision problems in management, business and economics. In particular, the following is an example how IFS bi-matrix games are applied to determining optimal strategies for e-commerce retailers.

With the rapid development of network, the competition between e-commerce retailers is becoming increasingly fierce. The high satisfaction degree of a customer can bring long-term profits and reduce the cost of attracting new customers for e-commerce retailers. Thus, how to improve the satisfaction degrees of customers has become a competitive target of e-commerce retailers.

Let us consider the case of two e-commerce retailers R_1 and R_2 (i.e., players I and II) making a decision aiming to enhance the satisfaction degrees of customers. As players' judgments for the satisfaction degrees of customers including preference and experience are often vague and players estimate them with their intuition. On the other hand, quite often it is easier to describe customers' negative feelings, i.e., their dissatisfaction degrees for e-commerce retailers (i.e., players). The IFSs can indicate customers' preference information in terms of satisfaction, dissatisfaction and neutralization. Thus, it is more realistic and appropriate to assume that payoffs of e-commerce retailers R_1 and R_2 (i.e., players I and II) are expressed with IFSs.

In this example, assume that e-commerce retailers R_1 and R_2 are rational, i.e., they will choose optimal strategies to maximize their own profits without cooperation. Suppose that retailer R_1 has three pure strategies: improving credit (α_1), establishing a scientific and rational service system (α_2) and providing customers with satisfaction products (α_3). Retailer R_2 possesses the same pure strategies as retailer R_1, i.e., the options of retailer R_2 are: improving credit (β_1), establishing a scientific and rational service system (β_2) and providing customers with satisfaction products (β_3).

Let us consider the following specific IFS bi-matrix game for this scenario, where the payoff matrices of e-commerce retailers R_1 and R_2 are expressed with IFSs as follows:

$$\tilde{A} = \begin{pmatrix} \{<(\alpha_1,\beta_1),0.90,0.05>\} & \{<(\alpha_1,\beta_2),0.70,0.20>\} & \{<(\alpha_1,\beta_3),0.50,0.40>\} \\ \{<(\alpha_2,\beta_1),0.40,0.40>\} & \{<(\alpha_2,\beta_2),0.60,0.15>\} & \{<(\alpha_2,\beta_3),0.70,0.10>\} \\ \{<(\alpha_3,\beta_1),0.50,0.40>\} & \{<(\alpha_3,\beta_2),0.90,0.10>\} & \{<(\alpha_3,\beta_3),0.60,0.30>\} \end{pmatrix}$$

and

$$\tilde{B} = \begin{pmatrix} \{<(\alpha_1,\beta_1),0.85,0.10>\} & \{<(\alpha_1,\beta_2),0.90,0.05>\} & \{<(\alpha_1,\beta_3),0.50,0.20>\} \\ \{<(\alpha_2,\beta_1),0.60,0.40>\} & \{<(\alpha_2,\beta_2),0.80,0.05>\} & \{<(\alpha_2,\beta_3),0.70,0.20>\} \\ \{<(\alpha_3,\beta_1),0.50,0.40>\} & \{<(\alpha_3,\beta_2),0.10,0.75>\} & \{<(\alpha_3,\beta_3),0.90,0.05>\} \end{pmatrix},$$

respectively, where the IFS $\{<(\alpha_1,\beta_1),0.90,0.05>\}$ in the payoff matrix $\tilde{A}$ means that the satisfaction (or membership) degree of customers is 0.9 and the dissatisfaction (or nonmembership) degree of customers is 0.05 for e-commerce retailer R_1 if he/she adopts the pure strategy α_1 and e-commerce retailer R_2 also adopts the pure strategy β_1. Other entries in the IFS payoff matrices $\tilde{A}$ and $\tilde{B}$ can be similarly explained.

Using Eq. (13), the parametric bilinear programming model is constructed as follows:

$$\max\left\{\begin{array}{l} (1.75+0.05\lambda_1+0.05\lambda_2)y_1z_1+(1.6+0.1\lambda_1+0.05\lambda_2)y_1z_2+(1+0.1\lambda_1+0.3\lambda_2)y_1z_3 \\ +(1+0.2\lambda_1+0\lambda_2)y_2z_1+(1.4+0.25\lambda_1+0.15\lambda_2)y_2z_2+(1.4+0.2\lambda_1+0.1\lambda_2)y_2z_3 \\ +(1+0.1\lambda_1+0.1\lambda_2)y_3z_1+(1+0\lambda_1+0.15\lambda_2)y_3z_2+(1.5+0.1\lambda_1+0.05\lambda_2)y_3z_3-u(\lambda_1)-v(\lambda_2) \end{array}\right\}$$

$$s.t.\left\{\begin{array}{l} (0.9+0.05\lambda_1)z_1+(0.7+0.1\lambda_1)z_2+(0.5+0.1\lambda_1)z_3\le u(\lambda_1) \\ (0.4+0.2\lambda_1)z_1+(0.6+0.25\lambda_1)z_2+(0.7+0.2\lambda_1)z_3\le u(\lambda_1) \\ (0.5+0.1\lambda_1)z_1+0.9z_2+(0.6+0.1\lambda_1)z_3\le u(\lambda_1) \\ (0.85+0.05\lambda_2)y_1+0.6y_2+(0.5+0.1\lambda_2)y_3\le v(\lambda_2) \\ (0.9+0.05\lambda_2)y_1+(0.8+0.15\lambda_2)y_2+(0.1+0.15\lambda_2)y_3\le v(\lambda_2) \\ (0.5+0.3\lambda_2)y_1+(0.7+0.1\lambda_2)y_2+(0.9+0.05\lambda_2)y_3\le v(\lambda_2) \\ y_1+y_2+y_3=1 \\ z_1+z_2+z_3=1 \\ u(\lambda_1)\ge 0, v(\lambda_2)\ge 0 \\ y_i\ge 0 \quad (i=1,2,3) \\ z_j\ge 0 \quad (j=1,2,3). \end{array}\right. \tag{16}$$

For $\lambda_1=0$ and $\lambda_2=0$, Eq. (16) becomes the bilinear programming model as follows:

$$
\text{ma}\left\{\begin{array}{l}1.75y_1z_1+1.6y_1z_2+y_1z_3+y_2z_1+1.4y_2z_2+1.4y_2z_3\\ +\ y_3z_1+y_3z_2+1.5y_3z_3-\mu_{\tilde{u}(\bar{x},\bar{y})}-\mu_{\tilde{v}(\bar{x},\bar{y})}\end{array}\right\}
$$

$$
s.t.\begin{cases}0.9z_1+0.7z_2+0.5z_3\le\mu_{\tilde{u}(\bar{x},\bar{y})}\\ 0.4z_1+0.6z_2+0.7z_3\le\mu_{\tilde{u}(\bar{x},\bar{y})}\\ 0.5z_1+0.9z_2+0.6z_3\le\mu_{\tilde{u}(\bar{x},\bar{y})}\\ 0.85y_1+0.6y_2+0.5y_3\le\mu_{\tilde{v}(\bar{x},\bar{y})}\\ 0.9y_1+0.8y_2+0.1y_3\le\mu_{\tilde{v}(\bar{x},\bar{y})}\\ 0.5y_1+0.7y_2+0.9y_3\le\mu_{\tilde{v}(\bar{x},\bar{y})}\\ y_1+y_2+y_3=1\\ z_1+z_2+z_3=1\\ 0\le\mu_{\tilde{u}(\bar{x},\bar{y})}\le 1\\ 0\le\mu_{\tilde{v}(\bar{x},\bar{y})}\le 1\\ y_i\ge 0\,(i=1,2,3),z_j\ge 0\;(j=1,2,3).\end{cases}\tag{17}
$$

Solving Eq. (17), we can obtain the optimal solution $(\boldsymbol{y}^*,\boldsymbol{z}^*,\mu_{\tilde{u}^*(\bar{x}^*,\bar{y}^*)},\mu_{\tilde{v}^*(\bar{x}^*,\bar{y}^*)})$, where $\boldsymbol{y}^*=(0.381,0.381,0.238)^{\mathrm{T}}$, $\boldsymbol{z}^*=(0.227,0.137,0.636)^{\mathrm{T}}$, $\mu_{\tilde{u}^*(\bar{x}^*,\bar{y}^*)}=0.618$ and $\mu_{\tilde{v}^*(\bar{x}^*,\bar{y}^*)}=0.671$. Thus, we obtain the satisfaction (or membership) degrees 0.618 and 0.671 of the satisfying Nash equilibrium values and corresponding satisfying Nash equilibrium (mixed) strategies $\boldsymbol{y}^*=(0.381,0.381,0.238)^{\mathrm{T}}$ and $\boldsymbol{z}^*=(0.227,0.137,0.636)^{\mathrm{T}}$ of e-commerce retailers R_1 and R_2, respectively.

Similarly, for $\lambda_1=1$ and $\lambda_2=1$, Eq. (16) becomes the bilinear programming model as follows:

$$
\max\left\{\begin{array}{l}1.85y_1z_1+1.75y_1z_2+1.4y_1z_3+1.2y_2z_1+1.8y_2z_2+1.7y_2z_3\\ +1.2y_3z_1+1.15y_3z_2+1.65y_3z_3-(2-\upsilon_{\tilde{u}(\bar{x},\bar{y})}-\upsilon_{\tilde{v}(\bar{x},\bar{y})})\end{array}\right\}
$$

$$
s.t.\begin{cases}0.95z_1+0.8z_2+0.6z_3\le 1-\upsilon_{\tilde{u}(\bar{x},\bar{y})}\\ 0.6z_1+0.85z_2+0.9z_3\le 1-\upsilon_{\tilde{u}(\bar{x},\bar{y})}\\ 0.6z_1+0.9z_2+0.7z_3\le 1-\upsilon_{\tilde{u}(\bar{x},\bar{y})}\\ 0.9y_1+0.6y_2+0.6y_3\le 1-\upsilon_{\tilde{v}(\bar{x},\bar{y})}\\ 0.95y_1+0.95y_2+0.25y_3\le 1-\upsilon_{\tilde{v}(\bar{x},\bar{y})}\\ 0.8y_1+0.8y_2+0.95y_3\le 1-\upsilon_{\tilde{v}(\bar{x},\bar{y})}\\ y_1+y_2+y_3=1\\ z_1+z_2+z_3=1\\ 0\le\upsilon_{\tilde{u}(\bar{x},\bar{y})}\le 1\\ 0\le\upsilon_{\tilde{v}(\bar{x},\bar{y})}\le 1\\ y_i\ge 0\,(i=1,2,3),z_j\ge 0\,(j=1,2,3).\end{cases}\tag{18}
$$

Solving Eq. (18), we obtain the optimal solution $(y'^*, z'^*, \upsilon_{\tilde{u}'^*(\bar{x}'^*,\bar{y}'^*)}, \upsilon_{\tilde{v}'^*(\bar{x}'^*,\bar{y}'^*)})$, where $y'^* = (0.755, 0.069, 0.176)^{\mathrm{T}}$, $z'^* = (0.222, 0.622, 0.156)^{\mathrm{T}}$, $\upsilon_{\tilde{u}'^*(\bar{x}'^*,\bar{y}'^*)} = 0.198$ and $\upsilon_{\tilde{v}'^*(\bar{x}'^*,\bar{y}'^*)} = 0.174$. Thus, we obtain the dissatisfaction (or nonmembership) degrees 0.198 and 0.174 of the satisfying Nash equilibrium values and corresponding satisfying Nash equilibrium (mixed) strategies $y'^* = (0.755, 0.069, 0.176)^{\mathrm{T}}$ and $z'^* = (0.222, 0.622, 0.156)^{\mathrm{T}}$ of e-commerce retailers R_1 and R_2, respectively.

According to the aforementioned discussion, the IFS of the satisfying Nash equilibrium value of e-commerce retailer R_1 is $\tilde{u}^*(\hat{x}^*, \hat{y}^*) = \{< (\hat{x}^*, \hat{y}^*), 0.618, 0.198 >\}$, which means that the satisfaction (or membership) degree of customers for e-commerce retailer R_1 is 0.618 when R_1 and R_2 choose the mixed strategies $y^* = (0.381, 0.381, 0.238)^{\mathrm{T}}$ and $z^* = (0.227, 0.137, 0.636)^{\mathrm{T}}$, and the dissatisfaction (or nonmembership) degree of customers for e-commerce retailer R_1 is 0.198 when R_1 and R_2 choose the mixed strategies $y'^* = (0.755, 0.069, 0.176)^{\mathrm{T}}$ and $z'^* = (0.222, 0.622, 0.156)^{\mathrm{T}}$, respectively. Similarly, the IFS of the satisfying Nash equilibrium value of the e-commerce retailer R_2 is $\tilde{v}^*(\hat{x}^*, \hat{y}^*) = \{< (\hat{x}^*, \hat{y}^*), 0.671, 0.174 >\}$, which means that the satisfaction (or membership) degree of customers for e-commerce retailer R_2 is 0.671 when R_1 and R_2 choose the mixed strategies $y^* = (0.381, 0.381, 0.238)^{\mathrm{T}}$ and $z^* = (0.227, 0.137, 0.636)^{\mathrm{T}}$, and the dissatisfaction (or nonmembership) degree of customers for e-commerce retailer R_2 is 0.174 when R_1 and R_2 choose the mixed strategies $y'^* = (0.755, 0.069, 0.176)^{\mathrm{T}}$ and $z'^* = (0.222, 0.622, 0.156)^{\mathrm{T}}$, respectively.

Furthermore, for different parameters $\lambda_1 \in [0, 1]$ and $\lambda_2 \in [0, 1]$, solving Eq. (16), we can obtain the satisfying Nash equilibrium values and corresponding satisfying Nash equilibrium strategies of e-commerce retailers R_1 and R_2 as in Tables 1–6, respectively.

Table 1. Satisfying Nash equilibrium values and corresponding strategies of e-commerce retailers

Parameters		R_1		R_2	
λ_1	λ_2	$y^{*\mathrm{T}}$	$u^*(\lambda_1)$	$z^{*\mathrm{T}}$	$v^*(\lambda_2)$
0	0	(0.381,0.381,0.238)	0.618	(0.227,0.137,0.636)	0.671
0	0.1	(0.411,0.348,0.241)	0.618	(0.227,0.137,0.636)	0.683
0	0.5	(0.541,0.223,0.236)	0.618	(0.227,0.137,0.636)	0.737
0	0.8	(0.659,0.132,0.209)	0.618	(0.227,0.137,0.636)	0.787
0	0.9	(0.705,0.101,0.194)	0.618	(0.227,0.137,0.636)	0.806
0	1	(0.755,0.069,0.176)	0.618	(0.227,0.137,0.636)	0.826

It can be easily seen from Tables 1, 2, 3 (or Tables 4, 5, 6) that the satisfying Nash equilibrium value of a player (i.e., I/e-commerce retailer or II/) only depends on his/her own preference/parameter regardless of other player's preference/parameter. However, strategy choice of a player is only affected by other player' preference/parameter.

Table 2. Satisfying Nash equilibrium values and corresponding strategies of e-commerce retailers

Parameters		R_1		R_2	
λ_1	λ_2	y^{*T}	$u^*(\lambda_1)$	z^{*T}	$v^*(\lambda_2)$
0.5	0	(0.381,0.381,0.238)	0.704	(0.243,0.312,0.445)	0.671
0.5	0.1	(0.411,0.348,0.241)	0.704	(0.243,0.312,0.445)	0.683
0.5	0.4	(0.507,0.253,0.240)	0.704	(0.243,0.312,0.445)	0.722
0.5	0.5	(0.541,0.223,0.236)	0.704	(0.243,0.312,0.445)	0.737
0.5	0.8	(0.659,0.132,0.209)	0.704	(0.243,0.312,0.445)	0.787
0.5	0.9	(0.705,0.101,0.194)	0.704	(0.243,0.312,0.445)	0.806
0.5	1	(0.755,0.069,0.176)	0.781	(0.232,0.542,0.226)	0.826

Table 3. Satisfying Nash equilibrium values and corresponding strategies of e-commerce retailers

Parameters		R_1		R_2	
λ_1	λ_2	y^{*T}	$u^*(\lambda_1)$	z^{*T}	$v^*(\lambda_2)$
0.9	0	(0.381,0.381,0.238)	0.781	(0.232,0.542,0.226)	0.671
0.9	0.1	(0.411,0.348,0.241)	0.781	(0.232,0.542,0.226)	0.683
0.9	0.4	(0.507,0.253,0.240)	0.781	(0.232,0.542,0.226)	0.722
0.9	0.5	(0.541,0.223,0.236)	0.781	(0.232,0.542,0.226)	0.737
0.9	0.8	(0.659,0.132,0.209)	0.781	(0.232,0.542,0.226)	0.787
0.9	0.9	(0.705,0.101,0.194)	0.781	(0.232,0.542,0.226)	0.806
0.9	1	(0.755,0.069,0.176)	0.781	(0.232,0.542,0.226)	0.826

Table 4. Satisfying Nash equilibrium values and corresponding strategies of e-commerce retailers

Parameters		R_1		R_2	
λ_1	λ_2	y^{*T}	$u^*(\lambda_1)$	z^{*T}	$v^*(\lambda_2)$
0	0	(0.381,0.381,0.238)	0.618	(0.227,0.137,0.636)	0.671
0.1	0	(0.381,0.381,0.238)	0.634	(0.232,0.165,0.603)	0.671
0.4	0	(0.381,0.381,0.238)	0.686	(0.242,0.269,0.489)	0.671
0.5	0	(0.381,0.381,0.238)	0.703	(0.243,0.312,0.445)	0.671
0.8	0	(0.381,0.381,0.238)	0.760	(0.238,0.473,0.289)	0.671
1	0	(0.381,0.381,0.238)	0.802	(0.222,0.622,0.156)	0.671

Table 5. Satisfying Nash equilibrium values and corresponding strategies of e-commerce retailers

Parameters		R_1		R_2	
λ_1	λ_2	y^{*T}	$u^*(\lambda_1)$	z^{*T}	$v^*(\lambda_2)$
0	0.5	(0.541,0.223,0.236)	0.618	(0.227,0.137,0.636)	0.734
0.1	0.5	(0.541,0.223,0.236)	0.634	(0.232,0.165,0.603)	0.734
0.4	0.5	(0.541,0.223,0.236)	0.686	(0.242,0.269,0.489)	0.734
0.5	0.5	(0.541,0.223,0.236)	0.703	(0.243,0.312,0.445)	0.734
0.8	0.5	(0.541,0.223,0.236)	0.760	(0.238,0.473,0.289)	0.734
1	0.5	(0.541,0.223,0.236)	0.802	(0.222,0.622,0.156)	0.734

Table 6. Satisfying Nash equilibrium values and corresponding strategies of e-commerce retailers

Parameters		R_1		R_2	
λ_1	λ_2	y^{*T}	$u^*(\lambda_1)$	z^{*T}	$v^*(\lambda_2)$
0	0.9	(0.705,0.101,0.194)	0.618	(0.227,0.137,0.636)	0.806
0.1	0.9	(0.705,0.101,0.194)	0.634	(0.232,0.165,0.603)	0.806
0.4	0.9	(0.705,0.101,0.194)	0.686	(0.242,0.269,0.489)	0.806
0.5	0.9	(0.705,0.101,0.194)	0.703	(0.243,0.312,0.445)	0.806
0.8	0.9	(0.705,0.101,0.194)	0.760	(0.238,0.473,0.289)	0.806
1	0.9	(0.705,0.101,0.194)	0.802	(0.222,0.622,0.156)	0.806

5 Conclusion

In some situations, determining payoffs of bi-matrix games precisely depends on players' judgments and intuition, which are often vague and not easy to be represented with crisp values and fuzzy sets. In the above, we model IFS bi-matrix games and develop the parametric bilinear programming models and method by using the defined order relation of IFSs given in this paper. The developed models and method may simplify the calculation of Nash equilibrium solutions of IFS bi-matrix games.

Furthermore, it is easy to see that the models and method proposed in this paper may be extended to IFS multiobjective bi-matrix games. And more effective methods of IFS bi-matrix games will be investigated in the near future. Also it is expected that the proposed models and method may be applied to solving many competitive decision problems in similar fields such as management, supply chain and advertising although they are illustrated with the example of the e-commerce retailers' strategy choice problem in this paper.

Acknowledgement. The authors would like to thank the associate editor and also appreciate the constructive suggestions from the anonymous referees. This research was supported by the key Program of National Natural Science Foundation of China (No. 71231003), the Natural Science Foundation of China (Nos. 71961004 and 71561008), the Science Foundation of Guangxi Province in China (Nos. 2012GXNSFAA053013, 2014GXNSFAA118010), the Post-doctoral Science Foundation of China (No. 2013M540372) and the Post-Doctor Fund of Shanghai City in China (No. 13R21414700). The Innovation Project of Guet Graduate Education (No. 2019YCX S082).

References

1. Zadeh, L.A.: Fuzzy sets. Inform. Control **8**, 338–353 (1965)
2. Vijay, V., Chandra, S., Bector, C.R.: Bi-matrix games with fuzzy goals and fuzzy payoffs. Fuzzy Optim. Decis. Making **3**, 327–344 (2004)
3. Maeda, T.: On characterization of equilibrium strategy of the bi-matrix game with fuzzy payoffs. J. Math. Anal. Appl. **251**, 885–896 (2000)
4. Bector, C.R., Chandra, S.: Fuzzy Mathematical Programming and Fuzzy Matrix Games. Springer, Berlin (2005). https://doi.org/10.1007/3-540-32371-6
5. Larbani, M.: Solving bi-matrix games with fuzzy payoffs by introducing Nature as a third player. Fuzzy Sets Syst. **160**, 657–666 (2009)
6. Atanassov, K.T.: Intuitionistic fuzzy sets. Fuzzy Sets Syst. **20**, 87–96 (1986)
7. Atanassov, K.T.: Intuitionistic Fuzzy Sets. Springer, Heidelberg (1999). https://doi.org/10.1007/978-3-7908-1870-3
8. Li, D.F., Cheng, C.T.: New similarity measures of intuitionistic fuzzy sets and application to pattern recognitions. Pattern Recogn. Lett. **23**, 221–225 (2002)
9. Li, D.F.: Multiattribute decision making models and methods using intuitionistic fuzzy sets. J. Comput. Syst. Sci. **70**, 73–85 (2005)
10. Li, D.F., Nan, J.X.: A nonlinear programming approach to matrix games with payoffs of Atanassov's intuitionistic fuzzy sets. Int. J. Uncertain. Fuzziness Knowl. Based Syst. **17**(4), 585–607 (2009)
11. Li, D.F.: Mathematical-programming approach to matrix games with payoffs represented by Atanassov's interval-valued intuitionistic fuzzy sets. IEEE Trans. Fuzzy Syst. **18**(6), 1112–1128 (2010)
12. Nan, J.X., Li, D.F., Zhang, M.J.: A lexicographic method for matrix games with payoffs of triangular intuitionistic fuzzy numbers. Int. J. Comput. Intell. Syst. **3**(3), 280–289 (2010)
13. Aggarwal, A., Mehra, A., Chandra, S.: Application of linear programming with I-fuzzy sets to matrix games with I-fuzzy goals. Fuzzy Optim. Decis. Making **11**, 465–480 (2012)
14. Nayak, P.K., Pal, M.: Intuitionistic fuzzy bi-matrix games. NIFS **13**, 1–10 (2007)
15. Li, D.F.: Extension of the LINMAP for multiattribute decision making under Atanassov's intuitionistic fuzzy environment. Fuzzy Optim. Decis. Making **7**(1), 17–34 (2008)
16. Li, D.F.: Extension principles for interval-valued intuitionistic fuzzy sets and algebraic operations. Fuzzy Optim. Decis. Making **10**(1), 45–58 (2011)
17. Zhang, X.M., Xu, Z.S.: A new method for ranking intuitionistic fuzzy values and its application in multi-attribute decision making. Fuzzy Optim. Decis. Making **11**, 135–146 (2012)
18. Owen, G.: Game Theory, 2nd edn. Academic Press, New York (1982)
19. Nash, J.F.: Equilibrium points in n-person games. Proc. Natl. Acad. Sci. U.S.A. **36**, 48–49 (1950)

20. Mangasarian, O.L., Stone, H.: Two-person nonzero-sum games and quadratic programming. J. Math. Anal. Appl. **9**, 348–355 (1964)
21. Lmeke, C.E., Howson, J.T.: Equilibrium points of bi-matrix games. SIAM J. Appl. Math. **12**, 413–423 (1964)
22. Yager, R.R.: Some aspects of intuitionistic fuzzy sets. Fuzzy Optim. Decis. Making **8**, 67–90 (2009)
23. Li, D.F.: Decision and Game Theory in Management with Intuitionistic Fuzzy Sets. Springer, Heidelberg (2014). https://doi.org/10.1007/978-3-642-40712-3

Author Index

Printed in the United States
By Bookmasters